Avinashkumar Vinodkumar Karre
Piping and Instrumentation Diagram

Also of interest

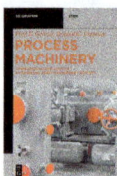

Process Machinery.
Commissioning and Startup: An Essential Asset Management Activity
Fred K. Geitner, Ronald G. Eierman, 2022
ISBN 978-3-11-070097-8, e-ISBN (PDF) 978-3-11-070107-4,
e-ISBN (EPUB) 978-3-11-070113-5

Process Engineering.
Adressing the Gap between Study and Chemical Industry
Michael Kleiber, 2020
ISBN 978-3-11-065764-7, e-ISBN (PDF) 978-3-11-065768-5,
e-ISBN (EPUB) 978-3-11-065807-1

Compressor Technology Advances.
Beyond 2020
Hurlel Elliott, Heinz Bloch, 2021
ISBN 978-3-11-067873-4, e-ISBN (PDF) 978-3-11-067876-5,
e-ISBN (EPUB) 978-3-11-067883-3

Sustainable Process Engineering
Gyorgy Szekely, 2021
ISBN 978-3-11-071712-9, e-ISBN (PDF) 978-3-11-071713-6,
e-ISBN (EPUB) 978-3-11-071730-3

Fluid Machinery.
Life Extension of Pumps, Gas Compressors and Drivers
Heinz Bloch, 2020
ISBN 978-3-11-067413-2, e-ISBN (PDF) 978-3-11-067415-6,
e-ISBN (EPUB) 978-3-11-067427-9

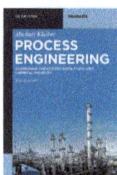

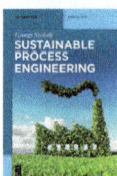

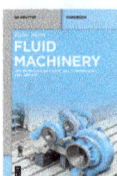

Avinashkumar Vinodkumar Karre

Piping and Instrumentation Diagram

A Stepwise Approach

DE GRUYTER

Author
Avinashkumar Vinodkumar Karre
Worley Group Inc.
4949 Essen Lane
Baton Rouge 70809
USA
avinash.karre@gmail.com

Contents in this book are solely based on the author's extensive work experience and knowledge. If part of the book or some contents matches the external source, it would be considered merely a coincidence.

ISBN 978-1-5015-1984-0
e-ISBN (PDF) 978-1-5015-1986-4
e-ISBN (EPUB) 978-1-5015-1335-0

Library of Congress Control Number: 2023935769

Bibliographic information published by the Deutsche Nationalbibliothek
The Deutsche Nationalbibliothek lists this publication in the Deutsche Nationalbibliografie;
detailed bibliographic data are available on the Internet at http://dnb.dnb.de.

Acknowledgments

I am thankful to my wife Vishali Tumma and my five-year-old son Arjun Karre for supporting my journey to complete this book. I am grateful to the entire De Gruyter team for helping me reach the finish line. Also, thanks to the Worley Group, who taught me everything about P&IDs.

I hope this book brings joy to the readers, and they can utilize it effectively on ongoing assignments.

Cheers!!!

https://doi.org/10.1515/9781501519864-202

Contents

Chapter 4
Control system components —— 48

Chapter 5
Process controllers, alarms, and safety instruments —— 62

Chapter 6
Utilities —— 66

Chapter 1
Understanding the basics of piping and instrumentation diagrams (P&IDs)

In this chapter, the introduction, importance, skeleton, comparison with block flow, process flow diagram, different tools, common rules, types, and steps involved in developing piping and instrumentation diagrams (P&IDs) are discussed in detail. Understanding the P&IDs is the first step in the development of critical equipment, instruments, and pipelines on the P&IDs. Once there is a good understanding of the basics of P&ID, the engineer can contribute efficiently towards the scope development of a project or participate in the P&ID meetings.

1.1 Introduction to P&ID

A Piping and instrumentation diagram (P&ID) is a detailed representation of a process showing piping and valves, instrumentation, process equipment, and note details required to design, construct, and safely operate a process plant. P&IDs are control documents for any process plant. P&IDs are commonly used by oil refineries, chemical plants, water treatment plants, food, power plants, oil & gas, engineering service companies, etc. P&ID is a medium used to communicate engineering information easily with others. Figures 1.1(a) and (b) show a process plant and a P&ID, respectively. From Figure 1.1(a), it can be seen that a processing plant has several processing units. Each process plant is different from the others. Each processing unit consists of multiple P&IDs. A processing unit can be easily understood by a set of P&IDs.

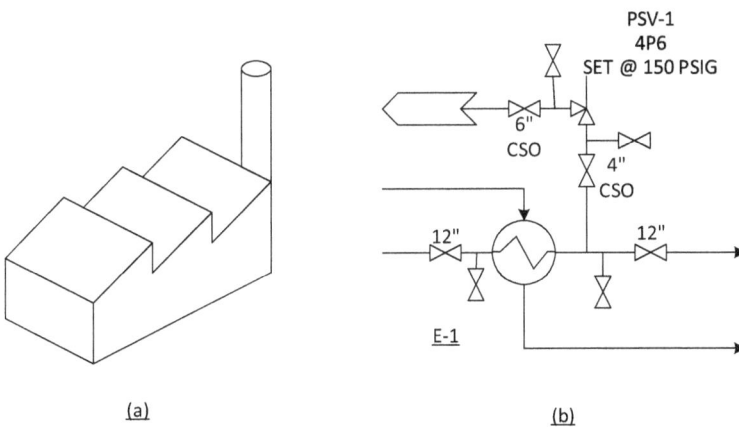

Figure 1.1: (a) A schematic of a process plant; (b) a schematic of a P&ID.

https://doi.org/10.1515/9781501519864-001

P&IDs are created by engineers who design a manufacturing process plant. The operating companies or the project or plant owners provide a contract to the engineering company. Typically, engineering companies develop or modify P&IDs based on the type of project, such as either grassroots or revamp. Oftentimes the technology licensor provides the P&IDs as a part of their basic engineering design package. Regardless of where the P&IDs are developed, the following key discipline engineers contribute mainly to developing the P&IDs:
- Chemical engineer
- Mechanical engineer
- Piping Design engineer
- Instrumentation engineer
- Plant operating engineer

There are several names for P&IDs practiced by many organizations and industrial plants. Some of the commonly used names are:
- Engineering Flow diagrams (EFDs)
- Engineering Line Diagrams (ELDs)
- Mechanical Flow Diagrams (MFDs)
- Utility Flow Diagrams (UFDs)

P&IDs are used by field technicians, plant operators, contractors, and engineers to understand the process. The plant operators or engineers use them for training new hires or trainees. Field technicians and plant operators use P&IDs for planning routine, maintenance, and shutdown tasks. Typical shutdown activities involve preparing a section of the process equipment or a process plant for pressure or leak test. Different activities involved in the maintenance tasks can be tracked using P&IDs. All routine or breakdown maintenance work permits are accompanied by highlighted P&IDs of the system being worked on, to show the scope of work. Engineers in engineering companies use P&IDs for revamp projects to debottleneck the existing plant and perform engineering studies. Engineers in operating companies use P&IDs to define the preliminary scope and make changes as per process requirements. P&IDs are also helpful in training new operators or technicians. New operators are usually given rigorous training inside a classroom before exposing them to the actual process plant and equipment.

1.2 Importance of P&IDs

Several functions are affected by the design and development of P&IDs. Some of these functions are control philosophy, equipment design, 3D model reviews, safety equipment requirements, etc. P&ID is used as a communication tool that helps in connecting with other disciplines. Figure 1.2 shows a schematic of communication involved in the development of P&IDs. A chemical engineer uses P&IDs to ask an equipment size-related question to a mechanical engineer, or a piping designer could ask about a free draining requirement of a piping from a chemical engineer.

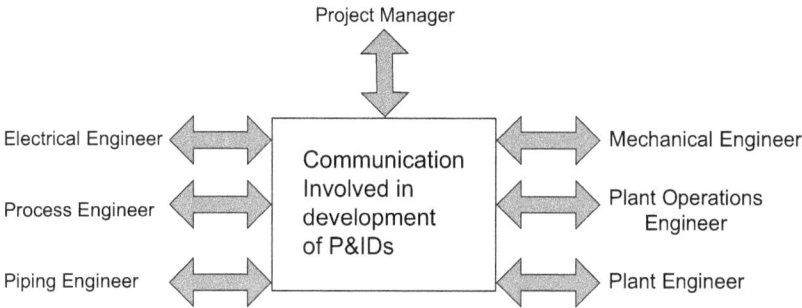

Project Manager

Electrical Engineer

Process Engineer

Piping Engineer

Communication Involved in development of P&IDs

Mechanical Engineer

Plant Operations Engineer

Plant Engineer

Figure 1.2: Communication involved in the development of P&IDs.

Piping designers use P&IDs to develop 3D piping models. P&IDs show essential elements such as piping sizes, equipment connections, isolation valves, drain valves, etc., much needed to develop 3D models. Piping designers also look for any special notes on P&IDs regarding a piping arrangement or equipment, e.g., designing a free draining of a pipe or locating a piece of equipment at 25 ft. above grade level. Such important notes help develop the 3D models. Also, piping designers use P&IDs for determining material take-off (MTO). MTO is the list of items such as piping size, length, number of gate valves, piping specification, etc. MTO is further used in cost estimation of the project. Piping designers use P&IDs to develop a line list. A line list document consists of all the engineering details for a pipe, such as operating and design temperature, operating and design pressure, pipe thickness, insulation type, etc. Also, once 3D models are reviewed and approved by a customer representative, the P&IDs are issued for construction. The piping designer uses P&IDs to develop piping isometric drawings. The piping contractors further use the isometric drawings of the piping for fabrication and installations.

Chemical engineers use P&IDs as a basis for a hazard and operability study (HAZOP). The chemical engineer marks a set of P&IDs for different sections of the process plant, and the HAZOP team, which usually consists of a plant operator, plant chemical engineer, chemical design engineer, HAZOP facilitator, and control system engineer, goes through different nodes and identifies hazards. Based on the identified hazards, appropriate design solutions are recommended and agreed upon by the team. These design solutions are further adopted either in the operating plant or in the ongoing and developing project.

P&IDs are helpful in planning hydrotesting and commissioning tasks of a processing plant. Different sections of the processing plant are divided into several sections based on the phase of the process fluid (gas or liquid phase). A section in the liquid phase is hydrotested using water, and a section in the gas phase is tested using high-pressure nitrogen. A processing plant is divided into several sections based on the steps involved in the commissioning process, such as water batching, start-up utilities, etc. Plant operators and engineers prepare a set of P&IDs for each section and divide responsibility for effectiveness. Figures 1.3 and 1.4 show the hydrotesting and commissioning steps in a schematic format.

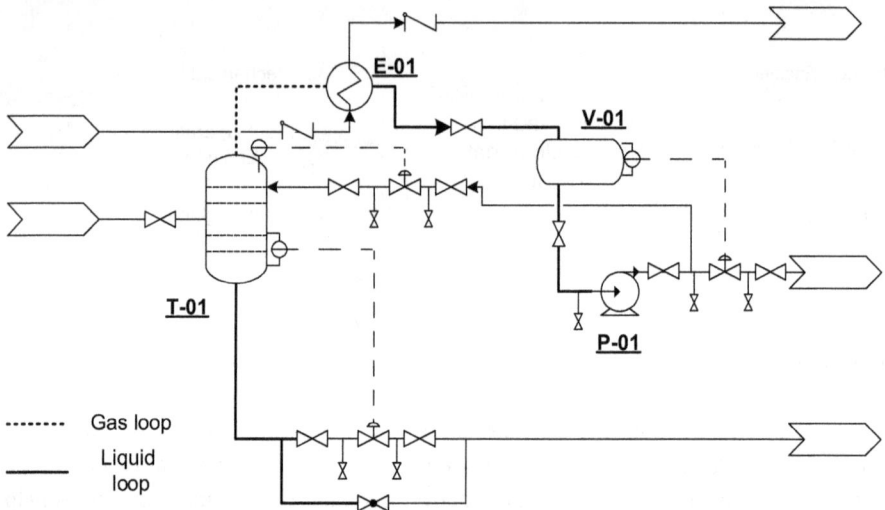

Figure 1.3: An example of a hydrotesting loop.

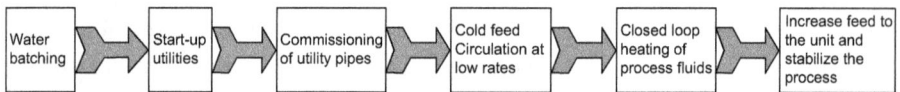

Figure 1.4: Steps involved in the commissioning of a processing unit.

P&ID helps instrumentation and electrical engineers develop engineering diagrams and logic narratives. P&IDs serve as a basis for developing a control loop programming and dynamic control system interface. P&IDs serve as a basis for preparing operating guidelines for the manufacturing plant and can be used for analyzing a safety incident.

The P&ID and the scope marked are developed through all the engineering phases. P&IDs are mainly used by chemical engineers in the engineering industry from initial scope through detail design development of the process plant. Adding a new scope on P&IDs involves showing a scope cloud around equipment, piping, and instrumentation. An example is shown in Figure 1.5.

As mentioned, there are several uses of P&IDs and many engineers and technicians use them in different scenarios. Figure 1.6 summarizes the importance of P&IDs in the industry.

P&ID consists of many design details such as equipment, line, specialty items, isolation valves, trips, alarms, and design notes, which are very significant to those who are

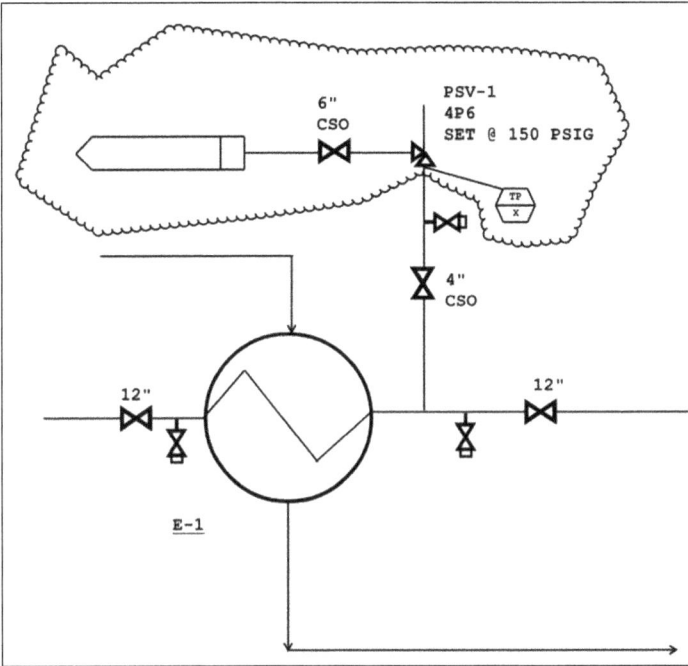

Figure 1.5: An example of scope mark-up on a P&ID.

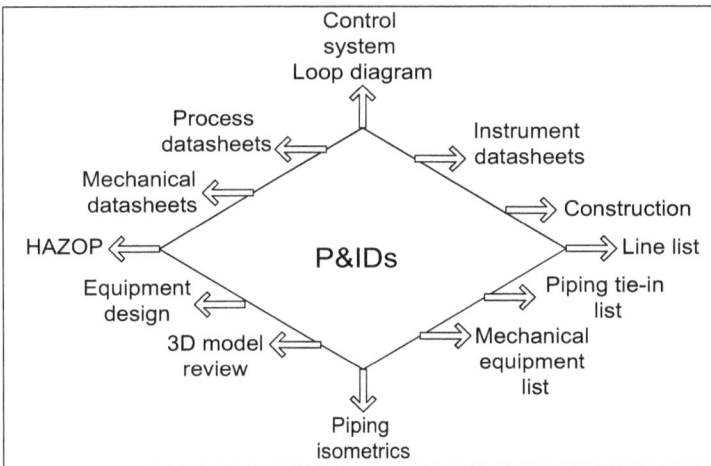

Figure 1.6: Importance of P&IDs.

studying the plant operations either for training or for a possible revamp project. One can find an equipment type and the design details of the equipment on the P&IDs.

There are several types of process pieces of equipment. The P&ID user can get an idea based on the type of equipment. Figure 1.7 shows a typical type of pump. The centrifugal pump is higher in capacity but lower in discharge pressure. On the other hand, positive displacement pumps are higher in discharge pressure but have lower capacity. The user can distinguish the type of equipment based on the symbology used. Figure 1.7(a) shows a centrifugal-type pump, and Figure 1.7(b) shows a positive displacement-type pump.

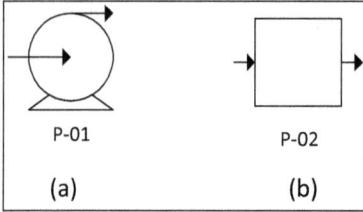

Figure 1.7: (a) Example of a centrifugal pump; (b) example of a positive displacement pump.

Design notes help understand the essential details about equipment or piping. Line information such as line size, service code, insulation thickness, piping specification, tracing type, etc., are helpful to the user in determining the nature and behavior of the flowing fluid inside the pipe. Isolation valves are useful in planning isolation of equipment or a section of a piping system. Figure 1.8 shows such example.

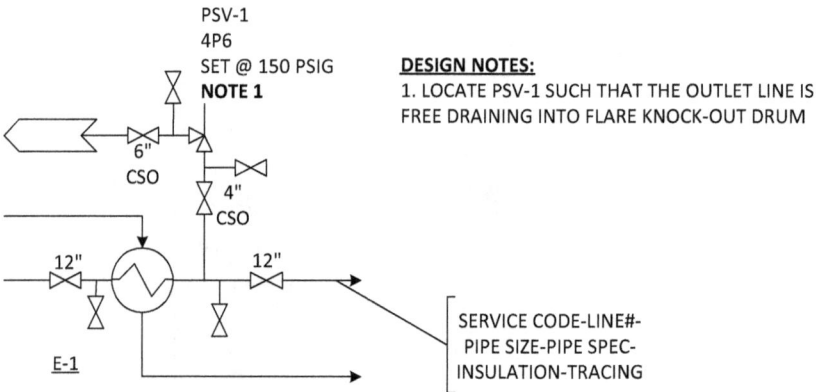

Figure 1.8: Importance of details on P&ID.

Understanding the difference between a board-mounted instrument and a locally mounted one is important. An alarm can be programmed on the board instrument, e.g., Figure 1.9(a) shows a high flow alarm when it reaches a set point. On the contrary, a locally mounted instrument shows local reading and can only be read by the field operators. Figure 1.9(b) shows an example of a locally mounted flow indicator.

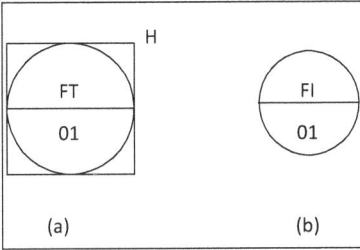

Figure 1.9: (a) A board-mounted flow instrument with an alarm function; (b) locally mounted flow instrument.

1.3 When to start developing P&IDs

The nature of P&ID development depends on a need for a new product or the modification purpose of the existing plant. When there is an idea is to make profits from the manufacturing plant, a conceptual block flow diagram (BFD) is initially developed. A simple process flow diagram (PFD) is developed, followed by complex and detailed P&IDs. If some modifications to the existing processing plant are required to comply with environmental regulations or to increase efficiency, marks are added to the P&IDs to reflect the scope. In the manufacturing unit, the need for modification to the existing P&IDs depends on the need for additional equipment to improve the plant's safety or to improve the efficiency of a running plant by modifications.

Typically, plant owners approach engineering contractors with a grassroots or a revamp project plan. New P&IDs are developed as a part of the grassroots project development either from scratch or with the help of an engineering licensor, depending on the process. Existing or as-built P&IDs are modified by engineering contractors depending on the revamp scope outlined by an operating company. In the engineering contractor's world, issued for an estimate (IFE) quality P&IDs in phase 2 are developed. Phases 3 and 4 are where P&IDs are further developed and finally issued for construction (IFC).

Figure 1.10 shows the initial steps involved in developing P&IDs. The process starts with someone's idea for chemical manufacturing or pharmaceutical industries. Funding is provided to perform laboratory-scale experimental work to check the idea's viability. Once the laboratory experiment results and viability are checked, a team of finance personnel and technical engineers further discuss the possible scale-up and profits. Once the project passes through this gate and when additional funding is received, a contract is typically given to an engineering contractor to start developing P&IDs and equipment design. If environmental regulations drives the project, the operating company has not much choice but to execute the project. Initially, the BFD is prepared to show the main scope, followed by PFD showing further details. BFD and PFD are then given to an engineering contractor to develop P&IDs. If the project type is based on capital efficiency to make profits, a similar process is followed, and typically these projects are cost- and schedule-driven and very challenging.

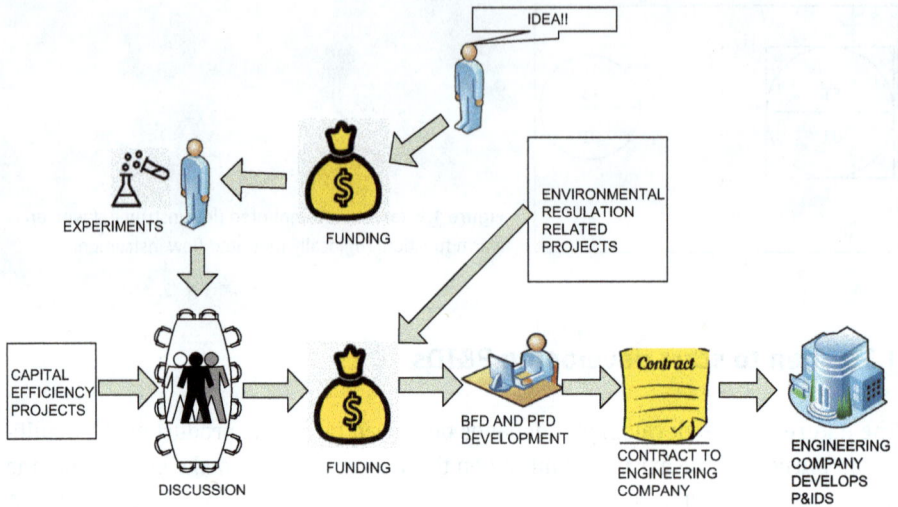

Figure 1.10: Initial steps involved in developing P&IDs.

1.4 Skeleton of a P&ID

Figure 1.11 shows a typical skeleton of a P&ID page. A P&ID usually goes through many revisions due to ongoing changes in the plant or a new project required for the plant. A revision block provides the change history carried out by previous users. Engineers from a new project can look up the detailed history based on the change history log, if they want to understand items in more detail. A revision block of a P&ID consists of the revision date, the name of the person who revised it, a short description of the change, the change number, etc. Table 1.1 shows an example of a typical revision block. Each revision is associated with a revision number, and a P&ID shows the change with this revision number. The comments section of the P&ID includes design, project, and hold notes, as shown in Figures 1.11 and 1.12.

The title of a P&ID is important. The title typically contains the owner's name and company's logo, the project number, a description of the P&ID, the P&ID number, and the current revision of the P&ID. This helps in tracking and shows the subject of the P&ID. Figure 1.13 shows a typical title of a P&ID.

1.5 Comparison of BFD, PFD, and P&IDs

A BFD is used for initial scope preparation, which needs to be approved by the customer before the next phase of the project kicks off. Once the project proposal is accepted, a PFD) is prepared, which adds more details regarding the type of equipment,

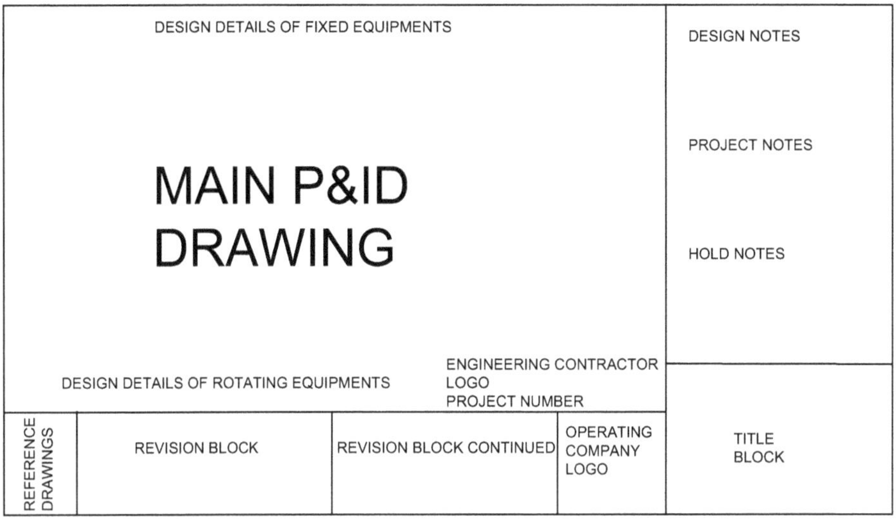

DESIGN DETAILS OF FIXED EQUIPMENTS			DESIGN NOTES
MAIN P&ID DRAWING			PROJECT NOTES
			HOLD NOTES
DESIGN DETAILS OF ROTATING EQUIPMENTS		ENGINEERING CONTRACTOR LOGO PROJECT NUMBER	
REFERENCE DRAWINGS	REVISION BLOCK	REVISION BLOCK CONTINUED	OPERATING COMPANY LOGO	TITLE BLOCK

Figure 1.11: Skeleton of a P&ID page.

Table 1.1: An example of a typical revision block.

Revision no.	Date	Marked by	Checked by	Approved by	Revision description
1	02/10/2020	AK	AB	AC	Issued for HAZOP
2	03/10/2020	AK	AB	AC	ADDED NEW P-01 A/B

Design notes:
1. SULFUR TRAP TO BE CONTRA-TRACED.
2. INTERNAL MIST ELIMINATOR IS REMOVED.

Project notes:
1. VALVE DETAILED TO BE PROVIDED BY VENDOR.
2. EXCHANGER E-01 IS MISSING FOUNDATION IN THE FIELD.

Hold notes:
1. PSV-01 SIZING TO BE CONFIRMED.
2. LINE 001 HYDRAULICS TO BE CONFIRMED

Figure 1.12: Typical comments section of the P&ID.

flow of materials, and connectivity of units. P&IDs take the PFDs and further add details such as line size, relief valve protection, valve sizes, etc. A simple comparison of BFD, PFD, and P&ID is shown in Figures 1.14(a)–(c), and Table 1.2.

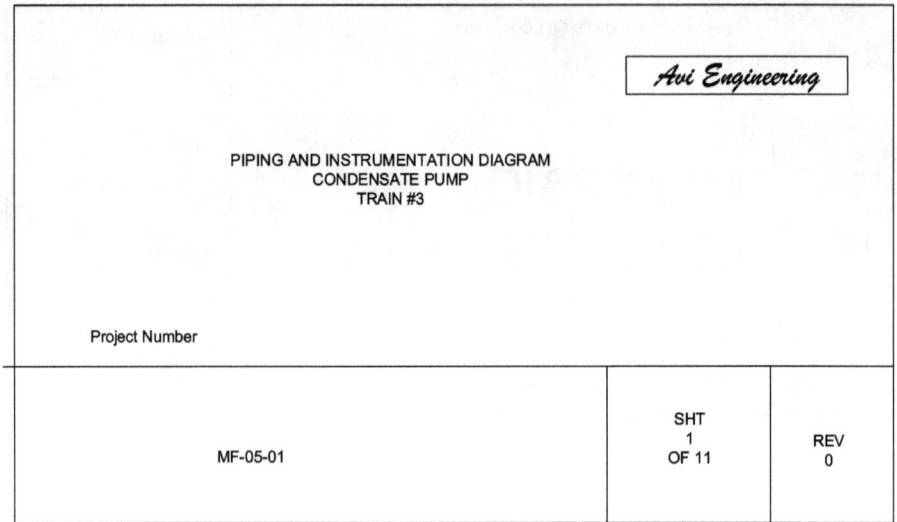

Figure 1.13: Typical title block of a P&ID.

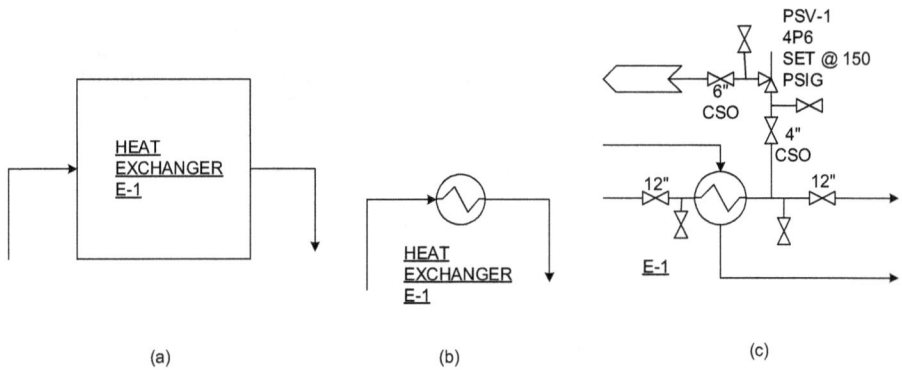

Figure 1.14: (a) Schematic of a BFD; (b) schematic of a PFD; (c) schematic of a P&ID.

1.6 Common tools used to develop P&IDs

Several tools are used to develop and draft P&IDs. In olden days, the engineer used D-size paper to add or delete marks to the P&IDs. Engineers from all disciplines only had a single master copy of D-size. It made the marking and tracking a bit difficult. At present, a Bluebeam Revu version of the software is widely used across the engineering industry and in operating plants.

There is flexibility in the Bluebeam software, whereby one can prepare a tool chest for a specific project. Once prepared, any engineer can pick and choose different

Table 1.2: Comparison of the PFD's and P&IDs.

Parameter	PFDs	P&IDs
Purpose	Show the concept of the unit	The detail diagram shows all the details that match the installed equipment, piping, and instruments in the unit
Complexity	Simple	Complicated and very detailed
Number of drawings for a unit	The process is shown in very few PFD drawings	A higher number of drawings are needed to show all the process and utility details
Equipment details	Just the equipment tag and description needed	Complete design details such as capacity, motor size, the material of construction, etc., of the equipment are needed
Piping details	Not needed	Complete piping design details such as line size, insulation, piping specification, service code, etc. are needed
Instrument tags	Instruments are typically not numbered	All instruments, including the instruments inside the loop, are numbered
Alarms	Not shown	Shown on all applicable instruments
Interlock details	Only the main interlock valves are shown	All interlock valves with closed-loop wiring details are shown
Control loops	Only a simplified loop is shown	Complete details of the transmitter, control loop, and element are shown

valves, pipes, and equipment symbology. The tool chest can be customized as per the customer's standard. The built-in tool chest allows engineers to save time and work efficiently as possible. All disciplines are consistent with the new marks on Bluebeam. Also, this allows conducting of P&ID review meetings very efficiently and saves time for customers and the engineering procurement and construction (EPC) industry. An example of such a custom tool chest is shown in Figure 1.15.

Depending on the complexity of the unit, different sessions can be added to the Bluebeam. These sessions can be shared with customers and EPC engineers. Since all these sessions are saved in the cloud, all the engineers can view, edit or make changes, make comments as required, and communicate with other engineers. An example of such a session is shown in Table 1.3. Typically, the P&ID drafting coordinator controls this session in the Bluebeam. The drafting coordinator can edit or add rights to the users as requested by the lead chemical engineer on the project.

Adobe is commonly used to develop initial conceptual redlines. Even though the products of Adobe and Bluebeam software look similar, Adobe software has a few limitations, for example, tracking the changes or lack of sessions, which are available in Bluebeam. Adobe software can also be used to quickly prepare a sketch of the idea

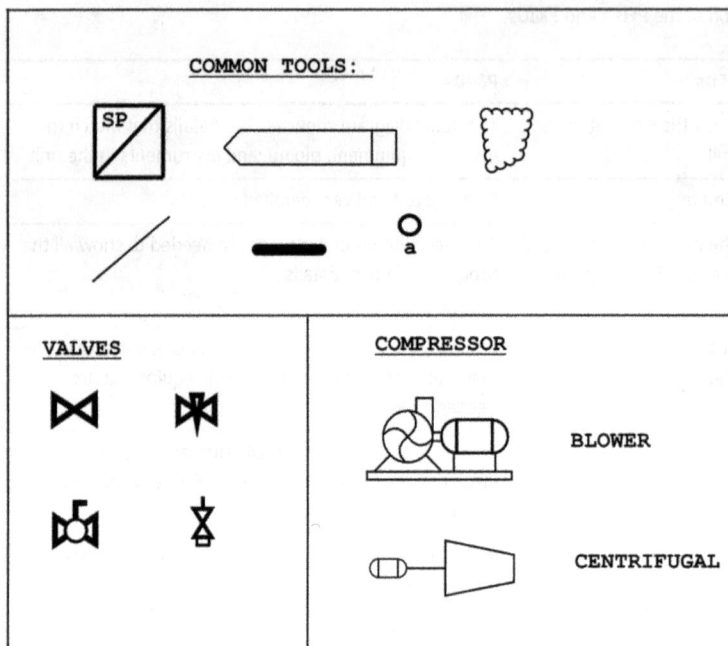

Figure 1.15: A common example of a few tools in Bluebeam.

Table 1.3: Different sessions in Bluebeam.

	Server name	
Initiate	Join the session	Leave the session
Session name and number	P&ID drawings	
Session name and number		
Session name and number		

of the process and communicate it with the customer in the preliminary stages of the project.

Once the P&ID is developed using the Bluebeam software, other software such as AutoCAD, MicroStation, and smartplant P&ID are used to draft the drawings. Depending on the customer's requirements, the EPC contractor may have to adapt to the drafting software. For example, the customer may prefer AutoCAD P&IDs because all their existing P&IDs are in AutoCAD. The new grassroots plant can take advantage of smartplant P&ID software. The smartplant P&ID software is interactive software where the information on the piping and equipment can be linked to the associated components on the P&ID. Also, the smartplant P&ID software speeds up the process of making the piping isometrics drawings. Overall, the smartplant P&ID software reduces the iterations required to draft the P&IDs, increasing the project's productivity.

1.7 General guidelines for developing P&IDs

There are several guidelines an engineer should keep in mind before developing P&IDs. Note that the general guidelines are not written in customer standards or anywhere else. They can only be understood if an engineer has sufficient knowledge of P&IDs. Below are the general rules:

One can use legend sheet to develop P&ID, as required. Legend sheets minimizes the mistakes that an engineer would typically make.

Using a good software such as Bluebeam can minimize the time required to prepare a P&ID, and an engineer can copy many templates from the tool chest menu, as discussed in Section 1.6. Creating a project-specific tool chest can make the P&ID development process efficient.

Lines on the P&ID should not cross each other as they represent the connection between two lines. The correct method is to draw a line, not cross it with the other line. Examples of such crossing lines are shown in Figures 1.16(a) and (b).

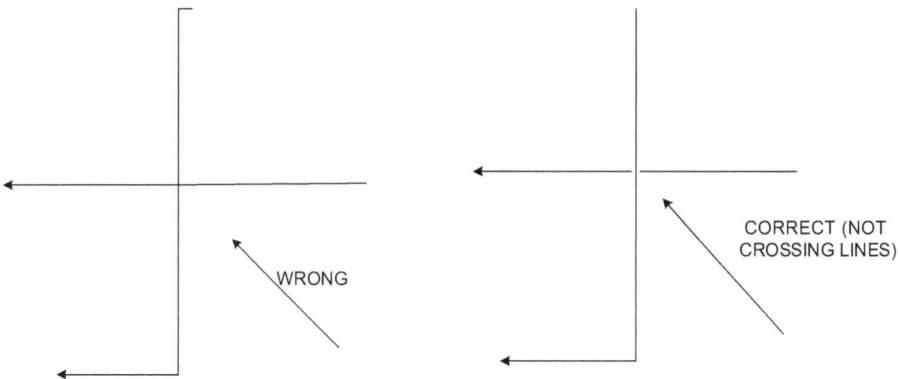

Figure 1.16: Example of crossing lines: (a) incorrect method; (b) correct method.

Different new equipment, piping, and instruments on P&IDs should be differentiated from existing as-built plant equipment, piping, and instruments. Also, after the marks are completed from the engineers, the P&IDs are drafted and back-checked. For these reasons, all EPC and customer companies follow different color representations of all items on P&IDs. Table 1.4 shows different color codes with their purpose.

All the associated equipment connected to the main equipment should be shown on the same P&ID page, if possible, for continuity. For example, a distillation column with a reboiler should be shown on the same P&ID page. Also, the equipment should be equally spaced on the P&ID page to accommodate other items such as piping and instrumentation details easily. Figure 1.17 shows an example of an equally spaced P&ID. Too many pieces of equipment on the same P&ID page could be a lot of clutter

Table 1.4: Color code representation for marking P&IDs.

Color	Purpose
Red	To add marks on the P&ID
Green	For deleting marks from the P&ID
Blue	To add drafting instruction
Light yellow highlighted	To back-check marks

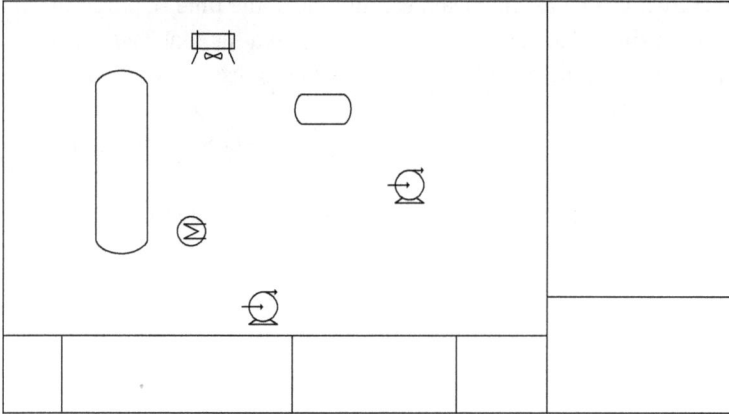

Figure 1.17: Example of an equally spaced P&ID.

and makes it challenging to convey the message to the design team. Figure 1.18 shows such cluttered P&ID.

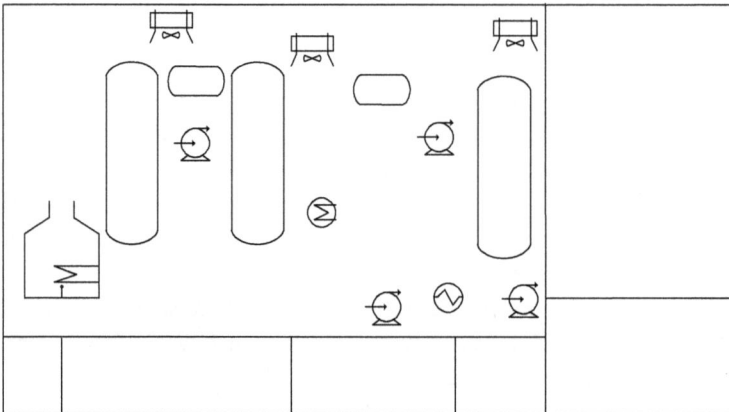

Figure 1.18: Example of cluttered P&ID.

The relative scale of equipment is an essential consideration while developing a P&ID. A distillation column is more significant in size than a pump and heat exchanger. The pump and exchanger should be drawn relatively smaller than the distillation column. Figure 1.19 shows an incorrect representation of the relative scale of different equipment.

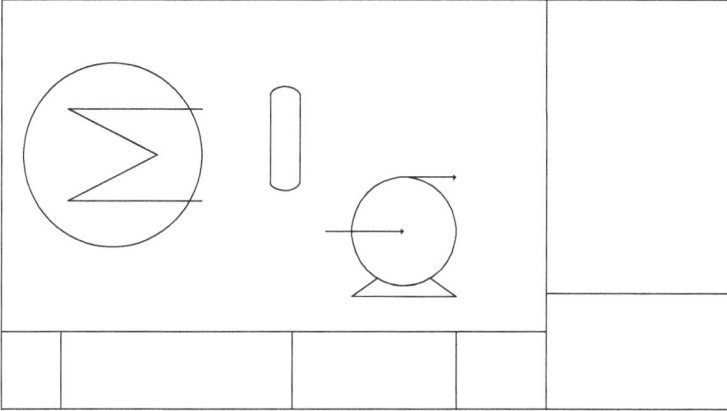

Figure 1.19: Example of the relative scale of the equipment.

All the necessary information, such as equipment, piping, and connectivity, should be defined in the early phases of the project to give a head start to other disciplines and help develop P&IDs efficiently in the further phases of the project. Having the block flow diagram and process flow diagram ready helps in developing P&IDs. Identifying hold items (insufficient information) early in the phases can help reduce rework for other disciplines. If the project is a grassroots expansion, the P&ID numbers and equipment numbers should be taken from customers early on to avoid rework later. If it is a very common piece of equipment such as a boiler, it is always better to start with the existing P&ID in order to save time, avoiding a lot of rework.

Off-page connectors are usually defined for process lines. Off-page connectors show the connectivity of lines from one page to other. For utility lines, a simple representation, such as a box with connecting utility P&ID is demonstrated. Figure 1.20 shows different versions of off-page connectors, depending on where the process stream is coming from, where it is going, and its utility.

1.8 Types of P&IDs

P&IDs are divided mainly into two categories. The first is process-related P&IDs. These consist of the main components of the process plant, such as distillation column, compressor, separation vessel, fired heater, pump, process piping, reactor,

Usually done to connect downstream process

Usually done when downstream process connects upstream

Usually done for utility connections

Some plant use numbers to designate utilities e.g. number 30 represents 50 psig steam

30

Usually done to connect downstream process

Usually done when line from the flowsheet connects upstream

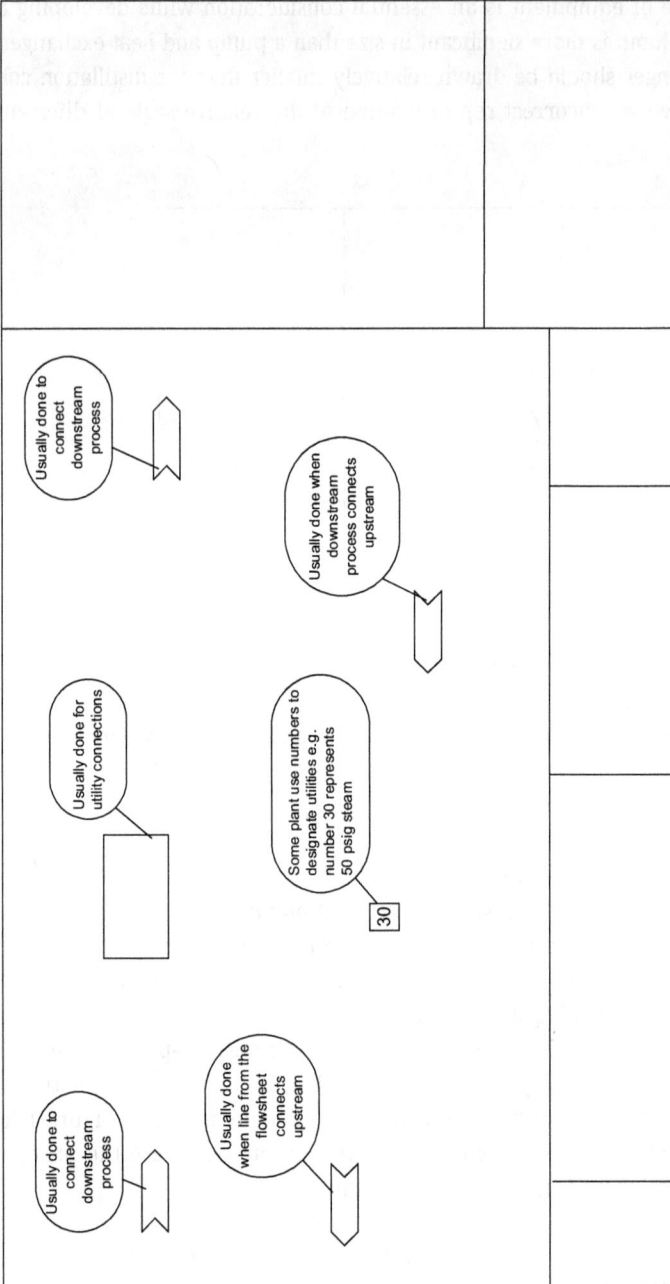

Figure 1.20: Off-page connectors and their meaning.

storage vessel, etc. The complexity of a set of P&IDs for a process plant depends on the number of process equipment, type of process, and type of process equipment. For example, P&IDs for caustic treatment of a propane process in the refinery is very simple and very few in equipment count.

On the other hand, P&IDs for a hydrocracker unit are very complex and have several equipment. The second category of the P&IDs is utility. The utility P&IDs show the distribution and production of several utilities such as cooling water, fuel gas, plant air, instrument air, utility water, demineralized water, steam, boiler feed water, etc. The utility distribution P&IDs show the connectivity of different utilities to all the process equipment and general utility requirements for the process plant. The utility production P&IDs show the actual production process of a particular utility. For example, steam production involves steam and mud drums, flue gas stack, air blower, burner arrangement, etc.

Tie-in and demolition drawings are required. They interface with the existing plant. These drawings are developed in conjunction with the project drawings. The demolition drawings show the actual demolition of a valve or piping for the scope. The tie-in drawings show where the actual piping or instrumentation tie-in is proposed by adding a new valve, piping, or instrumentation connection. Legend sheets are standard for different equipment, piping, instrumentation, etc. The legend sheet shows the customer standard for the development of P&ID. In every project, there are auxiliary P&IDs. The auxiliary systems, such as a list of equipment, pump or compressor seal plan, and sample station, are shown. For every P&ID, all the relief needs to be relieved into the flare disposal system. The flare and disposal system involves connecting network piping, the actual flare drum, and the flare. Figure 1.21 shows a broad representation of P&ID categories.

The bulk of the P&ID accounts for any process unit come from process and utility sections. Since they are critical for the process plant, it is important to understand their differences. Figures 1.22(a) and (b) show a simple differentiation between a process P&ID and a utility distribution P&ID. Process P&IDs connect different equipment through piping, valves, and instrumentation. Utility distribution P&IDs connect to the process P&IDs though piping and valves.

Vendor scope involves a third party providing the complete design of a section in the processing unit, e.g., vacuum jet ejector system. These vendor scope P&IDs are developed last as there are many unknowns during the normal development of a project. The vendor scope is typically shown by an empty box with a boundary and connections to and from the equipment. Figure 1.23 shows such an example of vendor scope P&ID.

The number of P&IDs required varies depending on the type of processing unit and type of P&ID. Table 1.5 shows such comparison. A simple process unit has a few P&IDs to represent a few pieces of equipment, and a complex process unit has several pieces of equipment that are shown on several P&IDs. Moreover, a complex process unit may require several process utilities to function, and they are shown on several utility P&IDs.

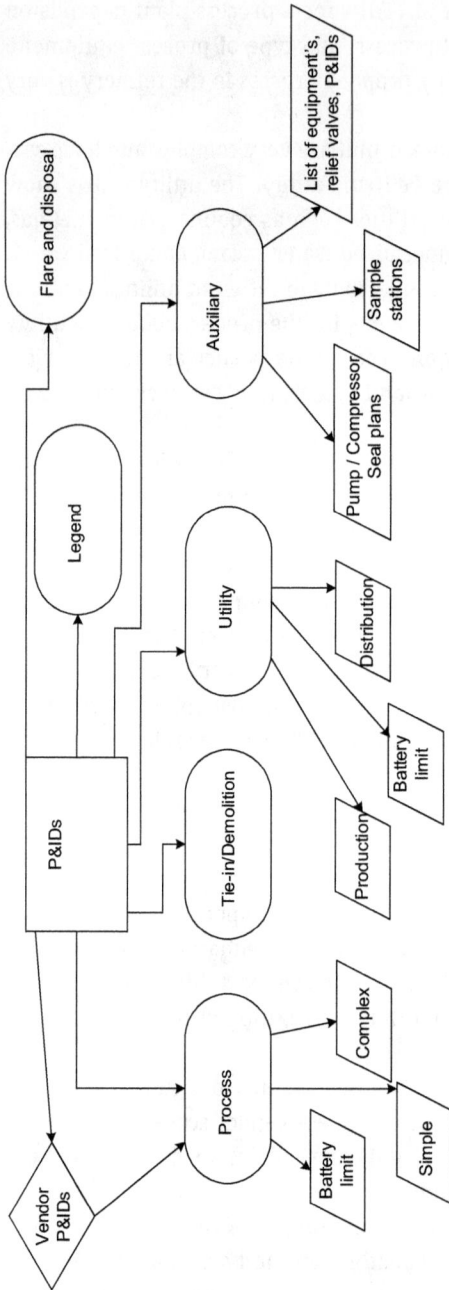

Figure 1.21: Broad representation of P&ID categories.

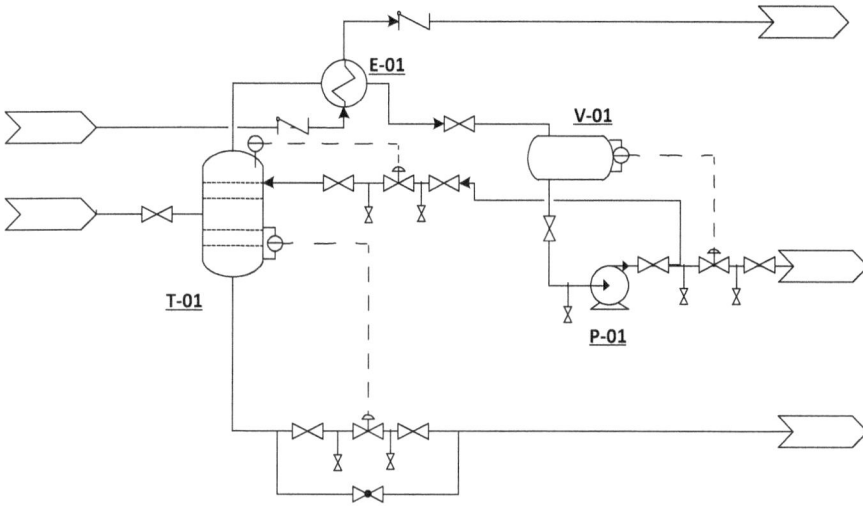

(a)

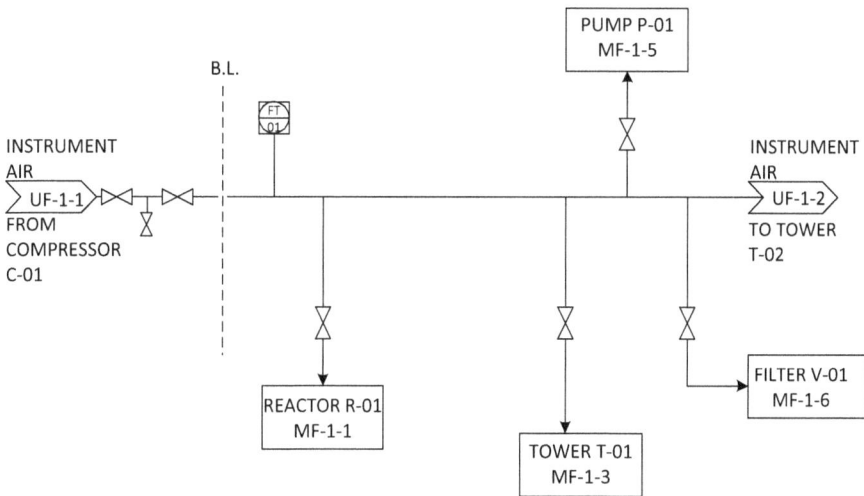

(b)

Figure 1.22: (a) A process P&ID; (b) a utility distribution P&ID.

Demolition of a process unit or piping section may be needed to accommodate a new piece of piping, valve, or equipment. Several lines show demolition at a 45° angle for the portion that needs to be removed. These drawings are named either DEMO-DRAWING NUMBER or D-DRAWING NUMBER. Once the demolition is completed, new project scope and valves are added and shown in the solid cloud. The tie-point (TP) represents the new piping tie-in needed to accommodate the new piping connection. Such an example of demolition and project P&IDs is shown in Figures 1.24(a) and (b).

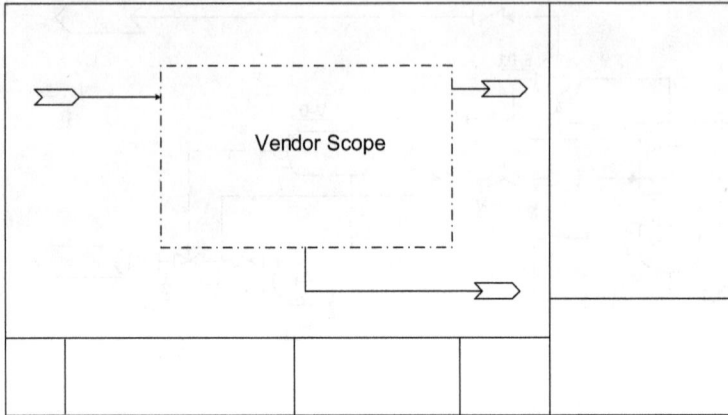

Figure 1.23: Vendor scope P&ID.

Table 1.5: Number of P&IDs per P&ID type and complexity of a process unit.

The complexity of process unit	P&ID type	Number of P&IDs
Simple	Simple (process)	Few
Complex	Complex (process)	Large in number
Complex	Battery limit (process)	Large in number
Simple	Battery limit (utility)	Few
Complex	Utility distribution	Larger in number

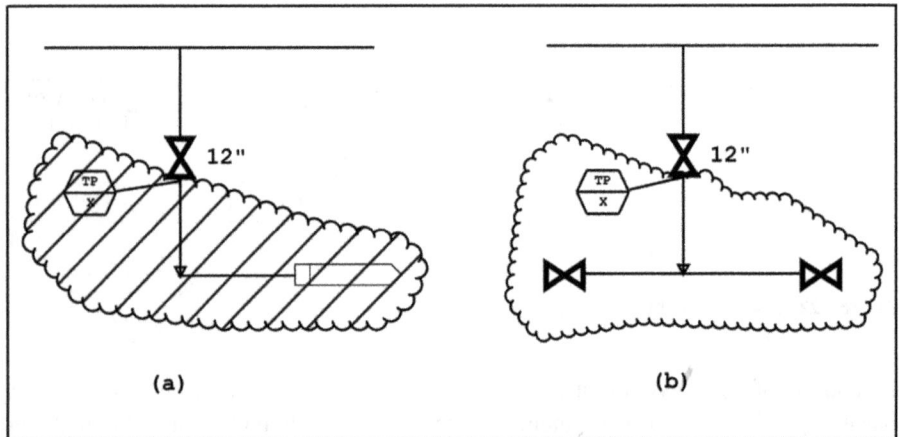

Figure 1.24: (a) Demolition P&ID; (b) project work P&ID.

1.9 Steps in developing P&IDs

Once the basics of P&IDs and the process of developing P&IDs are understood, further steps can be understood easily. If the initial step of understanding is not followed, the P&IDs can still be developed, but there would be some challenges in the development process. The first step in developing P&IDs is making sure the BFDs) and PFD's are ready because these two diagrams directly feed into the development of P&IDs. BFDs and PFDs can be used as a guide to review a number of equipment, the connections, the instrumentations, and the total scope of the project. The first step in developing P&IDs is to start planning for the number of equipment to be marked on a single P&ID. Once that is determined, the equipment's approximate location on that particular P&ID needs to be determined and drawn. When the equipment are marked, connect all the equipment and start adding details for the piping. Once that is done, add control system elements such as a control valve and offline instrument. Once the control system components are added, start adding the process controllers and instrumentation details. At this time, the P&ID is technically ready as a preliminary version. Now start adding required utility connections to P&ID. Now you have all the lines accounted for the P&ID; the next step is to add the details of isolation, drain valves, etc. Once that is done, additional information on the P&ID can be added, such as different types of notes. The same procedure can be repeated for other P&IDs in the project. Apart from the process and utility P&IDs, develop additional P&IDs for the plant safety system, as explained in the later section of this book. The entire process is shown in Figure 1.25.

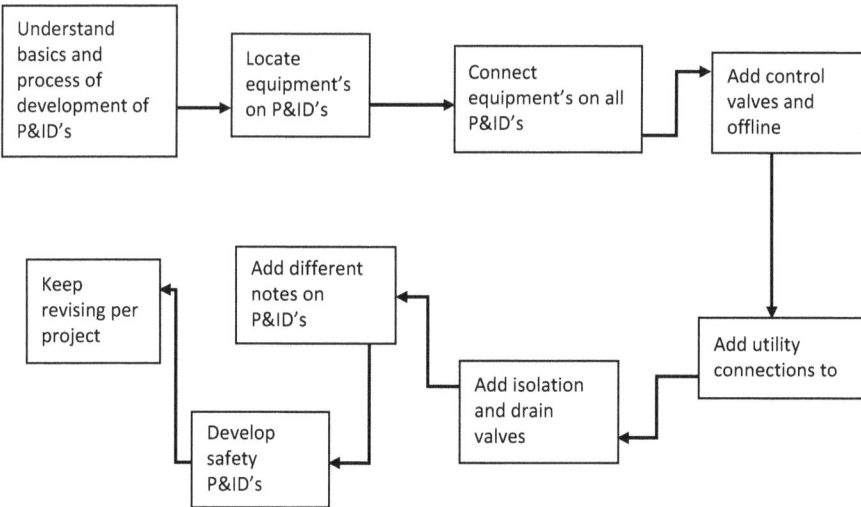

Figure 1.25: Block flow diagram for steps involved in the development of P&IDs.

Chapter 2
Development of equipment on P&IDs

Now that the fundamentals of P&ID are understood, the next step in developing a P&ID is identifying pieces of equipment in the process. A process engineer can start developing a P&ID from the BFD or PFD. PFDs as starting points are preferred because these are considered better than BFDs. The next step is about showing different equipment information and drawing equipment on the P&IDs. Equipment is a building block for the P&ID because everything either flows from or to the equipment. Depending on the process flow or design, various equipment are utilized in the process unit. Some equipment may be simple in nature e.g., a pump; some equipment maybe complex in nature, e.g., a distillation column. The complexity of the equipment affects the number of P&IDs the equipment may need to be spread or drawn. There are several process equipment in any given process. The type of equipment or the number of equipment depend on the type of process. A very complex process unit such as a hydrocracker unit can have a total count of 100 P&IDs with a total complex equipment count of 15 to 20.

2.1 Introduction to chemical engineering equipment and their types

Equipment is used in a process unit to transfer fluid, heat a cold fluid, cool a hot fluid, separate different components, store gas, solids, or liquid, reactions or conversion of a certain set of components into desired components, and disposal of waste. The equipment size is dependent on the capacity of the plant. Higher the capacity larger the size of the equipment, and vice versa. The size of the equipment is very important in determining the plot space required for the process unit. A typical revamp job is constrained by the plot space available. Process chemical engineers pay special attention to determining optimal size for the equipment. Process chemical engineers design the most commonly used equipment, such as drums, pumps, columns, and heat exchangers. Larger equipment are sized by specialized vendors. Examples of these special equipment are fired heater, reactors, flares, etc. Most of the time, a specialized vendor designs internals of a vessel or a column based on the process information provided by the process chemical engineer. Such internal designs include distillation column trays, vessel mist eliminators, etc. Several types of equipment are mainly defined by the type of operation, fluid phase, suitability of the fluid type, and the design required. The pumps can only be used in the liquid fluid service and compressors only in the gas service. A distillation column or series of columns are used when separating two or multicomponent systems. A pump or a compressor cannot be useful for

https://doi.org/10.1515/9781501519864-002

separation unit operation. A positive displacement pump can be used when very low flow and high discharge pressures are required.

On the contrary, centrifugal pumps are useful when very high flow capacities and low discharge pressures are needed. Figure 2.1 shows the type of equipment and Table 2.1 shows the purpose of different equipment. Static equipment do not have moving parts, and rotary equipment have moving parts, such as an impeller inside a pump.

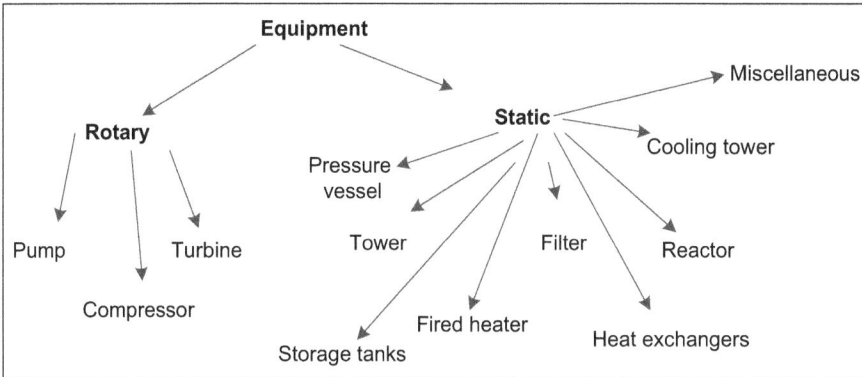

Figure 2.1: Types of equipment.

Table 2.1: Equipment type and the purpose.

Equipment	Use
Heat exchanger	Heat transfer between different fluids
Pump	Transfer of fluid from point A to B
Filter	Removes unwanted particles from a fluid stream
Tower	Separates two or multicomponent
Compressor	Compresses gas molecules to higher pressure
Fired heater	Transfers heat to the fluid
Reactor	Converts feed stream to desired products
Storage tank	Stores different fluids or solids
Cooling tower	Cools incoming hot water from the process
Turbine	Expands gas to lower pressure than suction
Pressure vessel	Stores liquid or gas at above atmospheric pressure

2.2 Space required on a P&ID for an equipment

P&ID is a critical engineering document used by various disciplines throughout the project and during operations. It is important to plan for the space required for the equipment on a P&ID. Leaving empty space or just showing one equipment on a single

P&ID means that the system will have a number of P&IDs for the unit, which could be troublesome for the users. Hence, it is important to optimize the space on the P&ID to include as many equipment as possible to show the process as clearly as possible. As a thumb rule, four to five equipment should be shown on a single P&ID. The idea behind the planning for space on a P&ID is to avoid clutter and ensure the users can follow the engineering diagrams clearly. The space requirement on a P&ID depends on the equipment type and complexity of the equipment, as shown in Tables 2.2 and 2.3.

Table 2.2: Complexity scale for determining the space requirement on a P&ID.

The complexity of an equipment	Complexity scale (1 to 5)	How much space is required on a P&ID
Simple	1	less than 10% of a P&ID page space
Simple to medium	2	15–25% of a P&ID page space
Medium to heavy	3	30–50% of a P&ID page space
Extra heavy	4	100% of a P&ID page space
Large	5	More than one P&ID page space

Table 2.3: Scaled equipment complexity for determining the space required on a P&ID.

	Scaled equipment complexity to determine space required on a P&ID
Heat exchanger	2
Pump	2
Filter	1
Tower	5
Compressor	5
Fired heater	5
Reactor	5
Storage tanks	4
Cooling tower	4
Turbine	4
Pressure vessel	2
Boilers	5

Equipment and their types need different spaces on P&ID. A column needs more space than the exchanger because a column has more inlet and outlet connections, valves, instruments, and relief valves. On the other hand, a simple exchanger has few piping, few valves, and one or two instruments, and a single relief valve.

Sometimes, the customer defines how many equipment can be present on a single P&ID. The EPC engineer should always consider the customer guidelines to avoid any future rework, which could be costly. For example, suppose the EPC engineer does not follow the customer guidelines at the beginning of the phase and specifies six heat exchangers and four pumps on a single P&ID. In that case, the project engineer notices that the EPC engineer did not follow the customer guidelines, for example, allowing only three exchangers and two pumps on a single P&ID. Now the EPC engineer must rework the previous diagram also affecting the connecting P&IDs, and carry out a lot of additional drafting work, which can be avoided.

The user or the engineer must consider that the other space would be taken up by piping, equipment design details, relief valve, controls, connectors, drawing names, project details, notes, revision blocks, etc.

Table 2.4 represents the number of P&IDs required depending on the type of equipment. From Table 2.4, the number of P&IDs required is the actual P&IDs

Table 2.4: Number of P&IDs required per the equipment type.

Equipment	The number of equipment's a single P&ID can accommodate	Number of P&IDs required for a single equipment
Heat exchanger	4–5	0.25
Pump	3–4	0.25
Filter	8–10	0.1
Tower	1 – simple 0.5 – medium complex 0.25 – very complex	1 – for simple 2 – medium complex 3–4 – very complex
Compressor*	0.5 – simple 0.2 – very complex	1–2 – for simple 4–5 – very complex
Fired heater	0.5	2
Reactor	1 – simple 0.5 – medium complex	1 – for simple 2 – medium complex
Storage tanks	1	0.5–1
Cooling tower*	1	1
Turbine	0.5–1	0.5–1
Pressure vessel	2	0.3–0.5
Boilers*	0.3	3

*The number of equipment or drawing count is based on auxiliary systems needed (e.g., chemical dosing system).

required for an equipment type. 0.25 indicates that the equipment occupies only 25% of a single P&ID.

The main operational and spare pumps are always shown on the same P&ID page as shown in Figure 2.2. Since it is difficult to fit the column drawing in a single P&ID page, it can be divided in half, as shown in Figure 2.3, and the other half can be shown on another P&ID page. This avoids clutter on a single P&ID page and allows the engineers to add details much more clearly, which is critical for the project.

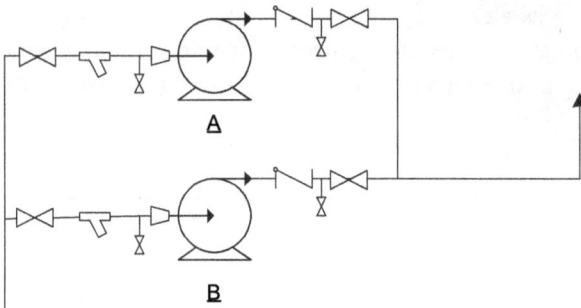

Figure 2.2: Representation of two pumps operating in parallel.

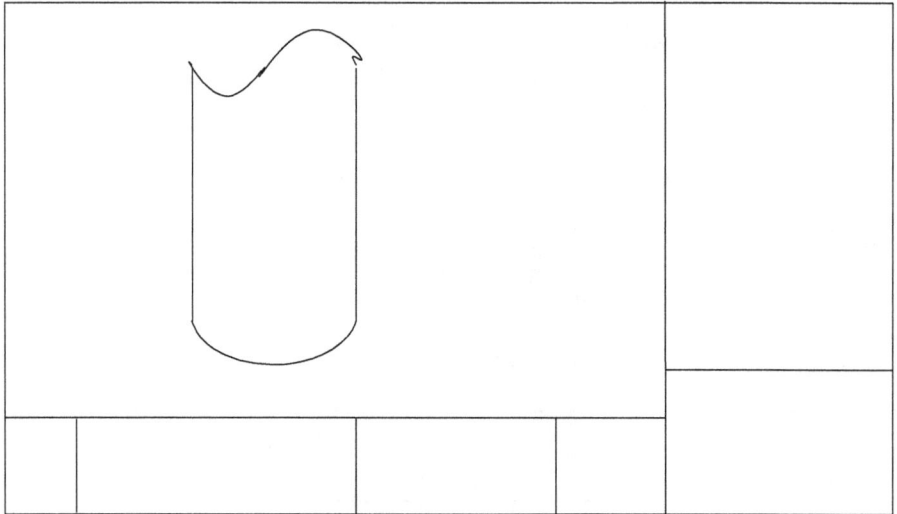

Figure 2.3: Representation of a distillation column on a single P&ID page.

Table 2.5 shows the actual count of process and utility area P&IDs for various units in a typical oil refinery. This table gives an idea to the users about how many P&IDs are expected from different units.

Table 2.5: Count of process and utility area P&IDs for various process units.

Unit operations	Description	No. of process P&IDs	No. of utility P&IDs
CDU	Crude distillation unit	25	10
VDU	Vacuum distillation unit	25	
FCCU	Fluid catalytic cracking unit	50	15
CRU	Catalytic reforming unit	25	10
ARU	Amine recovery unit	16	6
SRU	Sulfur recovery unit	20	6
TGTU	Tail gas treatment unit	10	3
SWS	Sour water stripper	6	2
SGU	Saturated gas unit	22	6
Deaerator	Used for the production of boiler feed water	–	4
Demineralized water	Used for the production of demineralized water	–	12
Steam	Used for the production of high- or medium-pressure steam (1 boiler)	–	4
Clarified water	Used for the production of utility water	–	4

2.3 Equipment design information on P&ID

Equipment information on P&IDs is necessary to understand design details about the equipment without looking into detailed documentation. Information on static equipment such as pressure vessels, columns, and tanks, is always located at the top of the P&ID. Information for rotary equipment, such as the pump and compressors, is located at the bottom of the P&ID. Figure 2.4 shows an example of rotary and static equipment design information on a P&ID.

Different information is needed as per equipment type. These are mentioned in the Table 2.6. Such design information is critical to the user of the P&ID. For example, a pump with a rate of 10 gallons per minute and 0.5 HP motor is considered a very small pump. Such information gives the user an idea of the pump's size. The other purpose of the design information is to get the equipment's construction material. During the consistency review of P&IDs, a review is important to know the metallurgy of equipment and the nearby piping. This information can be easily found on P&IDs. Table 2.6 shows details of some commonly used equipment in the industry.

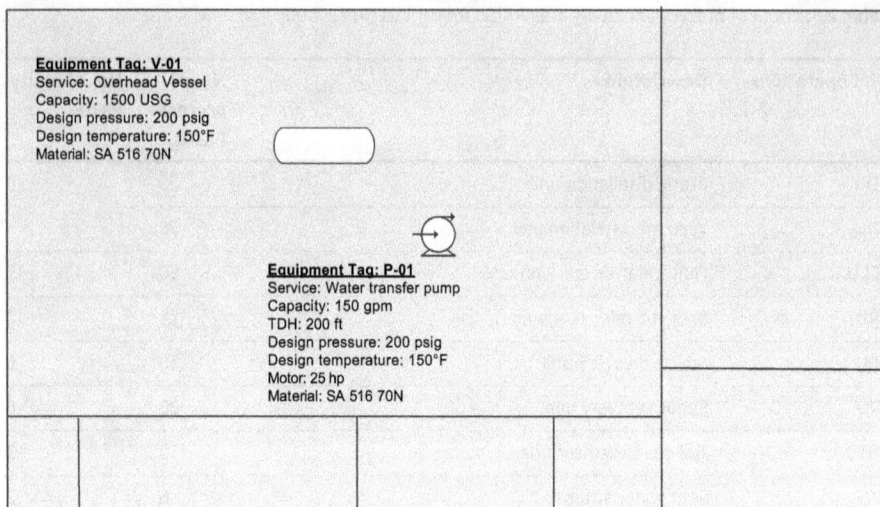

Figure 2.4: Design information on static and rotary equipment.

Table 2.6: P&ID details of commonly used equipment in the industry.

Equipment	P&ID details
Heat exchanger	Equipment tag: E-01 Service: Butane feed/bottoms exchanger Duty: 11 MMBtu/hr Material, Shell/Tube: C.S./C.S. Shell size: 4'-0" O.D. × 22'-0 length Shell design conditions: 250 psig @ 150 °F Tube design conditions: 250 psig @ 150 °F Exchanger type: AEU
Pump	Equipment tag: P-01 Name: Water transfer pump Capacity: 150 gpm TDH: 200 ft Design pressure: 200 psig Design temperature: 150 °F Motor: 25 hp Material: SA 516 70N
Filter	Equipment tag: F-01 Service: Feed filter Size: 28" O.D. × 110" Length Max DP: 20 psi Type of internals: 316 SS Design conditions: 250 psig @ 150 °F Material: C.S.

Table 2.6 (continued)

Equipment	P&ID details
Tower	Equipment tag: T-01 Service: Depropanizer tower Size: 3'-0" I.D. × 140'-0" T/T Material: C.S. Design conditions: 350 psig @ 250 °F Vacuum rating: FV/300 °F Number of trays: 50
Compressor	Equipment tag: C-01 Service: Propane compressor Motor: 2,500 hp Material: C.S. Capacity: 9,000 ACFM Discharge pressure: 280 psig Design conditions: 400 psig @ 450 °F
Fired heater	Equipment tag: H-01 Service: Hydrotreater feed heater Size: 40'-0" × 60-0" T/T Material: C.S. Duty: 40 MMBtu/hr Burners: 6 Design conditions: 600 psig @ 650 °F Vacuum rating: FV/300 °F
Reactor	Equipment tag: R-01 Service: Hydrotreater reactor Size: 6'-0" I.D. × 26-0" T/T Material: C.S. Design conditions: 600 psig @ 650 °F Vacuum rating: FV/300 °F
Storage tanks	Equipment tag: T-01 Service: Water storage tank Material: C.S. Size: 10'-0" O.D. × 17'-0" Height Capacity: 11,500 USG Design conditions: 2.5 psig @ 150 °F / 2 in WC
Cooling tower	Equipment tag: CT-01 Service: Cooling tower Duty: 160 MMBtu/hr Capacity: 28,000 gpm

Table 2.6 (continued)

Equipment	P&ID details
Pressure vessel	Equipment tag: V-01 Name: Water storage drum Size: 4'-0" ID × 10'-0" T/T Capacity: 1,000 USG Design pressure: 200 psig Design temperature: 150 °F Material: SA 516 70N

Equipment design information is also useful in sizing preliminary equipment during phase one. Most of the equipment information is available on equipment data sheets, but for a quick estimate in the first phase, this information can be used by engineers. In the preliminary phases of the project, a quick estimate of the volume or the functioning of the system can be estimated based on this information during the meetings, without looking at the actual equipment data sheets.

2.4 Pumps

A pump is a rotary device that transfers fluid from A to point B. The most common pumps are centrifugal and reciprocating or positive displacement types. Centrifugal pumps are designed to have a higher flow rate and lower head. On the flip side, the positive displacement pumps are designed to provide a very low flow rate at a high head.

The pump has an inlet and outlet isolation valve. The pump has an inlet strainer, which can be temporary or permanent. The pump discharge system also has a pressure gauge, or sometimes, a pressure transmitter to monitor the pressure in the outlet system. Depending on the service of the pump fluid, the pump can have a closed drain or an open drain. For example, propane and butane, which have high vapor pressure, cannot be drained into the close drain slop or sump, so this liquid must be depressurized into the vent system. The vent system is further connected to the flare system. Somewhat heavier hydrocarbons such as diesel or vacuum gas oil cannot be drained into the open drain system. These have to be drained into the closed drain system and collected into a slop system for further processing. The spectacle blinds are usually installed on the drain and vent lines to positively isolate the pump system from the vent and drain systems. For simple services such as water, the open drain is allowed. For the drain or vent purpose, a couple of 3/4-inch valves are installed on the pump's inlet and outlet and on the pump's casing.

Most pumps are used in chemical plants, refineries, oil and gas plant applications, water treatment units, pharmaceuticals, etc. Among all the pump types, centrifugal pumps are most commonly used in most applications in different plant areas. Positive displacement pumps such as chemical injection are used at very few locations where low volume flow rate is needed.

The driver for the pump can be an electric or steam turbine-driven motor. The small positive displacement pumps required for the driver are light-duty, and an electrically driven motor can be used. For centrifugal pumps, motor size is dictated by the head required, the specific gravity of the liquid, the pressure required at the destination, and system losses. Motor size can vary from 5 horsepower (hp) to 10,000 hp. Depending on the criticality of the pump, the motor type is chosen. For example, the feed pump to the reactor can be provided with a steam turbine-type motor for reliable operation of the plant. Without the feed pump, the reactor could fail, and the catalyst could be damaged. Steam turbine motors are more reliable and provide a constant power supply if there is an adequate steam supply.

Almost all centrifugal pumps have a minimum circulation line to protect the pump design from damage during turndown conditions. For example, the minimum flow for a pump is 50 gallons per minute (GPM), and the turndown condition is 25 GPM. In such a case, 25 GPM is the forward flow to the process, and 25 GPM is circulated back to the pump suction through the minimum circulation line to meet the minimum circulation requirement for the pump. For a very small pump, a fixed minimum circulation flow rate is designed using a restriction orifice.

Several alarms and switches are installed on a pump to remotely monitor or operate the pump. The motor (MO) is connected to the motor control center (MCC) room, and stop and start switches are provided with a run light. Figure 2.5 shows generic electrical symbols for the pump.

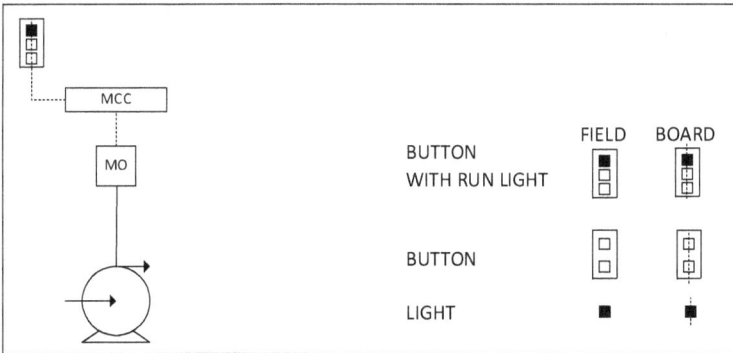

Figure 2.5: Motor symbology for the pump on the P&IDs.

Net positive suction head (NPSH) is a critical parameter for the pump design. If adequate NPSH is not provided, cavitation of the pump occurs, which will lead to damage of pump impeller and the pump itself. There are two terms related to NPSH: NPSH available and NPSH required. NPSH available is based on a few parameters such as the frictional losses in the pump suction piping, vapor pressure of the pumped liquid at an operating temperature, the suction vessel's height, the suction vessel's operating pressure, etc. NPSH required is the design parameter provided by the pump vendor. A process engineer must ensure that the NPSH available is always greater than the required NPSH, so that the pump can operate safely. It is common practice to elevate the suction vessel or the column to provide adequate NPSH.

Strainers and filters are usually provided in the pump section to protect the pump from any solid carryover from the suction vessel. Any solids or foreign matter can damage the impellers, eventually affecting the pump performance. For small pumps, strainers are usually employed in the suction. Strainer elements are usually installed during a startup as there is a high likelihood of solid carryover from the suction vessel. Once the startup is complete, the strainer element is removed and placed in the warehouse for further use. Oil degrades over time for pump services such as hot oil operating at very high temperatures. It is required to remove the degraded oil and solids from the closed loop system using filters. Spare filters are always installed to reduce maintenance time or to avoid disturbance to the running plant operations.

Vent and drains are installed on every pump. Those provide ease of cleaning during startup and maintenance activities. The type of drain depends on the process fluid service. The following are some drain types per the fluid type:
1. Normal liquid (water) drains – open drain, e.g., water
2. Hazardous liquid drain – closed drain to hazardous drain sump, e.g., dilute HCl
3. Oily drains – closed drain to the sump, e.g., kerosene
4. Gas vapor vents – flare or flare vapor recovery, e.g., LPG

The pump casing is designed for the maximum pressure the pump produces. It is possible that the suction piping of the pump is given a lower rating than the pump discharge pipe rating due to allowable pressures. Also, it is possible that the discharge fluid can travel back to the suction side. It is a good practice to rate the suction piping of the pump up to the valve to match the rating of the discharge piping.

Pipe rating depends on the upset pressure of this system. The suction pressure of the pump is mainly controlled by the pressure in the suction vessel, and the discharge pressure of the pump is mainly controlled by the pump design. Generally, the pump discharge pressure is higher than the pump suction pressure requiring a higher-rated pipe class than the pump suction piping.

It is crucial to keep the pump at the best operating range recommended by the pump vendor. If the pump operation deviates from the best operating range very frequently, it may damage the pump design over time. Several controllers are added to the pump discharge. If high flow is received in the pump suction vessel, this excess level is controlled by opening more flow from the pump discharge automated valve. Sometimes, the pump outlet control valve does not get a signal from the pump suction vessel, and it may be just a simple flow control valve. Adjusting the flow for the pump within recommended values of the pump design is important for such cases.

Variable frequency drive (VFD) adjusts the pump impeller's speed based on the system's pressure requirement. If the system's pressure requirement is lower, the pump needs to work less hard to produce less flow. If the pump system has a variety of flow and head cases, it is always recommended to add a VFD to save on power during normal process operations. Figure 2.6 shows a generic VFD control setup on a P&ID.

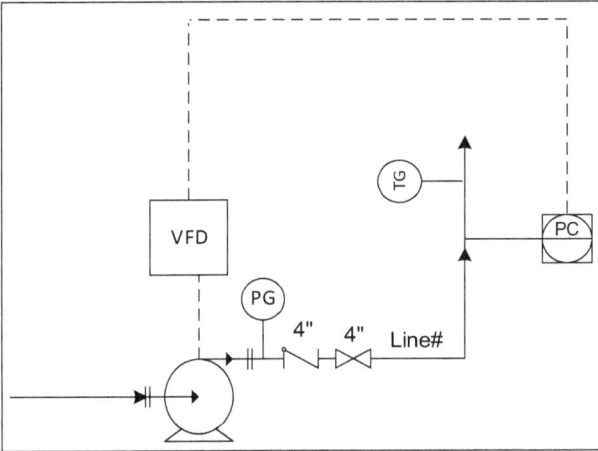

Figure 2.6: Pump with a VFD control.

A single suction valve and single isolation valve on a pump discharge are adequate. A check valve is also added on the pump discharge line before the isolation valve to ensure that there is no backflow. Single isolation on the suction and discharge is adequate for low-pressure systems. Double isolation with a bleed may be required for a high-pressure system to ensure safety. Double check valves are added on the pump discharge side, depending on the safety criticality of the system and customer guidelines.

The performance of a pump during the field walk is monitored using a locally mounted pressure gauge. A pressure instrument connected to the dynamic control system (DCS) or board is added to the common line of the pump downstream of the isolation valve to monitor the health of the pump remotely.

A flowmeter is always present on the pump discharge common line, which is helpful for adjusting the flow rate or monitoring the pump's performance.

Frequently, the pump vendor provides an internal relief valve for these small positive displacement pumps. For a positive displacement pump, a relief valve is provided on the pump discharge line, and the outlet of the relief valve is connected back to the pump suction or the suction vessel of the pump. The relief valve is set at a pressure within equipment design pressure of the pump discharge equipment.

Priming of the pump ensures no vapor molecules are trapped inside the pump. Otherwise, they could cavitate during startup. During pump startup for a centrifugal pump, it is a good practice to turn on the pump with the suction valve lined up and the discharge valve closed and open the drain valve on the pump discharge and bleed out all the vapors. Once the operator in the field confirms no further vapor, the pump's discharge line can be opened.

Since there is a differential pressure across a pump, it is always a good idea to take a sample or install a sample station across a pump. Having multiple sample stations in a unit for several services is a common practice.

There are two nozzles for every pump: one on pump suction and one on the pump discharge. Typically, the pump nozzle sizes are smaller than the pump suction and discharge pipe sizes, and reducers are always required on the pump suction and discharge sides. The flat bottom reducers are used for services with high vapor pressure or a tendency to produce vapor.

Sections 11.6 to 11.9 and 11.30 explain different pump systems, drains, safety, and valves.

2.5 Compressor and auxiliary

The compressor is complex machinery that compresses process gases in the processing plant. There are mainly two types of compressors: positive displacement and centrifugal compressors. Centrifugal compressors are used for most of the applications in the industry. They are designed for high volumes at the lower head. Positive displacement compressors are used for applications where low volumes are needed but at a very high head.

Centrifugal compressors generate heat when the gas is compressed. The heat is removed using air coolers. The centrifugal or positive displacement compressors cannot have liquid in the inlet. For these reasons, a suction vessel with the demister pad is added to remove any liquid from the inlet stream.

A compressor is one of the critical equipment in any process unit operation. To monitor the health of the compressor system, temperature, pressure, flow, etc., are monitored. A surge control mechanism is employed when the compressor receives flow much lower than the design value. Each compressor, similar to pumps, has several seal systems to alert the operator if there is any leak from the hazardous process. Buffer gas or nitrogen gas is used for the seal system.

2.6 Pressure vessel and internals

Pressure vessels provide either a storage capacity or perform a separation function. The vessels with simple storage capacity do not have any internals, and they could be feeding liquid to a pump. The vessels with a separating function could have more than two phases coming in and out. Based on the vessel's geometry and the internals, the vessel can separate vapor, aqueous, and organic phases.

The vessel size is dependent on two factors. One of the factors is the storage time needed. The storage time required could differ for different services; for example, the overhead vessel of the column needs about 15 to 20 min of residence time, while a vessel that feeds a heater or a reactor needs a residence time of at least 30 min. The other factor in designing a vessel is based on the degree or efficiency of separation needed. If the vapor or gas is involved in the feed, the internals inside the vessel must ensure that the gas is separated with the desired micron sizes of liquid in the gas phase. This can be critical for some services, such as a compressor suction knockout pot. Any liquid droplets greater than 150 micron in size can damage the compressor blades if not removed from the suction vessel. A demister pad is always added to improve the suction vessel's efficiency.

Different pressure vessel systems are discussed in detail in Sections 11.11 to 11.14.

2.7 Storage tanks

Various tank designs are available and can be used depending on the properties of the fluid. A fluid with very high vapor pressure, such as butane, needs a spherical tank, providing a uniform pressure across the entire sphere. Liquids with very low vapor pressure and are non-hazardous in nature, such as water, can be stored in the fixed roof tank. The fixed roof tank has a nonmoving roof, which can be used when there is no concern over vapor losses from the tank. Hazardous liquids with relatively low-vapor pressure, such as kerosene or heavy naphtha, can be stored in a floating roof tank. The floating roof tank provides a good mechanism that avoids any vapor leaking from the tank to the atmosphere. Liquefied natural gas (LNG) tanks are special in design because the temperatures in the tank are very low to keep the natural gas in liquefied form.

Every tank has one or more than one inlet. A product or a raw material is received through this inlet from a different unit or a pump. A pump is installed next to the tank, which can take the fluid and ships out to the rail or truck loading station or feeds it to the unit as a feedstock. A nozzle with the piping is connected to the pump suction. These off-site pumps often have a minimum circulation line that goes back to the tank. The tank also has two nozzles that are involved in heating the tank and removing the condensate or returning the heating fluid to the return header. Nitrogen blanketing may be necessary depending on the pressure control strategy on the tank. As the tank empties and fills up, the liquid level in the tank changes constantly and the vapor space also changes with the fluid level. To avoid changes in the pressure in the vapor space and to avoid contamination with air or water, nitrogen blanketing is done with the split range line going to the flare. A pressure vacuum valve that acts as the overpressure protection during filling and emptying events is also installed on the tank. Sometimes, the tank also needs a mixing mechanism to mix the fluids before they can be shipped to the end users. The common mixing mechanism is by adding an impeller attached to the motor or a couple of jet mixers or eductors.

Special precautions are needed for high vapor-pressure service tanks or spheres. During the external fire scenario, when the surrounding tanks are heated, the content in the tanks also heats up, which causes overpressure. This can lead to catastrophic failure and hazardous material can release into the atmosphere. To avoid releasing of hazardous chemicals, a water sprinkler arrangement or deluge safety system is installed around such tanks and spheres.

Often, the product stored in the tank receives some water along with the feed; over time, it builds up in the tank and settles at the bottom. If the water is not removed from the tank, the water may cause a serious operational problem in the downstream processing units. For example, water is not permitted in the high-temperature vacuum tower as it can flash immediately and cause damage to the column internals. To remove the water, a water bucket design is installed, and a provision is provided to remove any collected water at the bottom of the tank. Also, the tank's surface at the bottom is slightly sloped towards this water bucket, which can help collect water in the water bucket more efficiently. An example of the water bucket is shown in Figure 2.7.

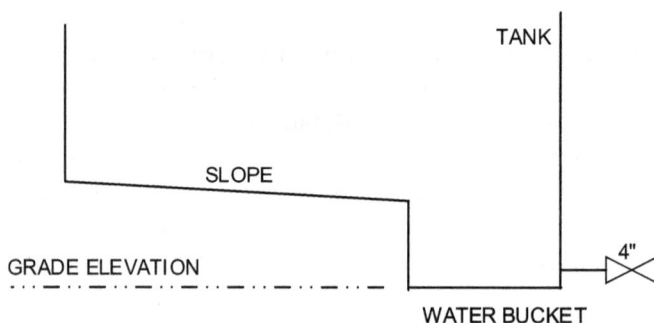

Figure 2.7: Water bucket setup for a tank.

Every tank has several instruments; the most common are level and temperature. Two different types of level indicators measure the level in the tank to provide flexibility in the design. The temperature instruments are located in the liquid region to measure the temperature of the stored liquid during the pump-out or filling event.

Sections 11.22 to 11.24 discuss different details of storage tanks.

2.8 Reactors and internals

Reactors are similar to pressure vessels, but they have either one or more than one catalyst systems inside them. The function of the catalyst is to change the chemistry of the incoming feed to make products as desired. The reactor size depends on the amount of catalyst to make a product. Every catalyst deactivates over time for various reasons, such as coking, fouling, metal poisoning, etc. Due to the deactivated catalyst, the reactor may not perform as per the original design, and towards the end of the life of the catalyst, a catalyst needs to be replaced. The old catalyst is removed from the reactor vessel using a nozzle, which is typically located at the bottom. The same nozzle can be used to load the fresh catalyst into the reactor bed.

The reactor has an inlet piping distributor to make sure the liquid or vapor that comes in is uniformly distributed across the catalyst bed, and there is no channeling happening, which could lead to bad performance of the reactor. Similarly, if there are multiple catalyst beds, there could be more than one distributor. At the bottom, catalyst guard screens are provided to prevent catalyst fines from escaping into the outlet piping.

The type of reactor depends on the type of reactions and the type of components involved in the process. The reactor can be a fixed bed, fluidized catalyst- type, continuous stir-type, or plug flow reactor. Depending on the exothermicity or endothermicity of the reaction, the reaction may need cooling or heating, provided by utilities externally or internally in the reactor.

As the performance of the reactor is a function of temperature and pressure, these are the commonly monitored parameters around the reactor bed. As there could be multiple beds in a single reactor and to ensure there is no channeling of the liquid or vapor, which can cause localized overheating or cooking, multiple temperature transmitters are added and monitored by the panel operator.

Reactors are considered pressure vessels and safety valve protection is needed to account for overpressure scenarios such as external fire, blocked outlet, abnormal heating scenario, runaway reaction, and addition of wrong material in the reactor.

Often, multiple reactors could be needed depending on the product rates and product specification and also the limitations of the first reactor. It may not be possible to show all the details of one reactor on a single page, and multiple pages of P&ID may be needed.

High temperature in the catalyst can deactivate the catalyst and can also form coke. To avoid such coking formation on the catalyst, safety instrumentation system is generally installed that measures different temperatures and compares them, and shuts the reactor down when one of the temperatures is very high t. Section 11.10 explains the reactor system with an example.

2.9 Heat exchanger

An exchanger is a simple device that exchanges heat or temperatures between two different fluids at different temperatures. An exchanger has inlet and outlet isolation valves on both sides . Also, drains and vents are installed on both sides of the exchanger, so that maintenance can be carried out during fouling or shutdown events.

Shell and tube-type heat exchangers are very commonly used in the industry because of their ease of maintenance and high heat transfer rates. Fin fan-type of exchanges are used mostly for vapor services where a large area of shell and tube heat exchanger might be needed. In addition to the larger area of the shell and tube heat exchangers, higher amounts of cooling water could be needed where fin fan exchangers are suitable. The plate and frame exchangers are used when the temperature approach required between two different fluids is very tight, and these types of exchanges can only be done in a very clean service fluid.

Shell and tube types of the reboiler can be attached to the column in vertical or horizontal arrangement to provide the necessary duty required for the column. The reboiler duty can be adjusted using a control valve provided on the hot side of the heat exchanger and by monitoring the target value of the tray temperature of the column.

The heat exchanger can be shown in two ways on the P&ID. One way is to show details of the channel side and shell side. Another type is to show a simple circle representing a shell side and the coil inside the circle that shows the tube side. Figure 2.8 shows different representations of a heat exchanger on a P&ID.

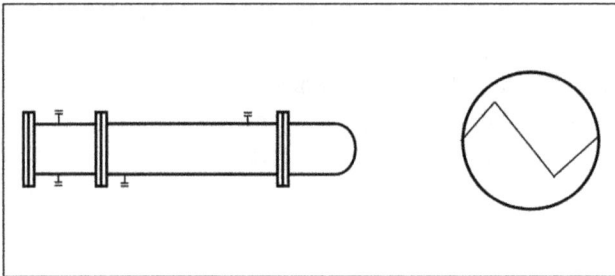

Figure 2.8: Representation of shell and tube heat exchanger on a P&ID.

Isolation valves and drain valves are installed on all sides of the heat exchanger to facilitate isolation and draining purposes during maintenance or shutdown time. To determine the performance of the heat exchanger from time to time, pressure and temperature transmitters and gauges are installed on all the lines of the exchanger. If the pressure drop across the exchanger is high, fouling is the likely reason, meaning that the exchanger needs cleaning very soon. Also, if the exchanger duty is dropped, the exchanger is not performing to its original design and may need review of potential problems such as fouling or leaking of tubes, etc.

Safety valves to protect the heat exchangers are often added to the piping connecting to the heat exchanger. The exchanger may need two safety valves to protect the hot side as well as the cold side. Note that the tube rupture scenario can be eliminated on the low-pressure side by designing the low-pressure side to have at least 10/13th value of the high-pressure side design pressure. But the exchanger still needs to be protected from other scenarios, such as a blocked outlet or external fire.

Suppose a fluid experiences a lot of changes in temperature, which would bring a lot of thermal expansion along the exchanger arrangement, an exchanger that has the flexibility to expand during temperature changes is chosen. On the cooling water side, a typical ¾" × 1" size safety valve size is selected for any thermal expansion scenarios.

Blinds are always installed on the heat exchanger nozzles to facilitate complete isolation of the exchanger for cleaning purposes.

A bypass around the exchanger is needed when there is excessive pressure drop across the exchanger or if the exchanger needs to be taken out for maintenance service.

Vapors are typically seen in the air coolers, and sometimes non-condensable vapors are encountered by the air coolers. As the non-condensable gases cannot condense at the temperature and pressure conditions of the air cooler, these gases need to be removed from the air fan (fin fan) header system. Typically, one inch-piping arrangement from the header box connected to the overhead vessel is done.

Sections 11.15 to 11.19 discuss different types of exchanger systems.

2.10 Columns

Distillation columns are used to separate two or multicomponent using a pressure vessel and trays or packing. Trays or packing provide the required mass transfer between vapor and liquid. To generate vapor, a reboiler or heater is used. If the heating is light-duty, a simple shell and tube exchanger can be used, and if the heater duty is heavy-duty, a fired heater needs to be used. The vapor generated from the tower needs to be condensed to get equilibrium in the tower. A fin fan or combination of a fin fan and a trim cooler is used to condense vapors from the tower. The condensed liquid is collected into the overhead vessel, part of the overhead liquid is pumped into the tower as reflux, and part of it is sent out to storage or to other processes for further processing.

Other types of towers include absorption, extraction, or simple wash column. All the columns have trays to provide a necessary mass transfer. Reboiling or condensation equipment depends on the amount of vapor the feed has and also the heat required to generate vapor from the feed stream.

Every tower has a feed nozzle, which can be a distribution pipe for effective distribution of the gas and vapor, and liquid distribution. There can be two nozzles at the column bottom, one for the liquid product pump out and the second for the reboiler inlet. Also, at the column bottom, one nozzle or pair of nozzles where the reboiler return gas and vapor mixture is returned. The column also has a vapor draw-off nozzle at the top and reflux for cooling the tower. These are some common nozzles that the distillation column has. Apart from these generic process nozzles, there are nozzles for pressure, temperature, and level transmitters. Note that for a complex column such as a crude column, there can be anywhere between 30 and 40 nozzles, which depends on multiple pump around, multiple draw-offs, and multiple injections of liquids inside the tower.

Every column needs routine cleanup or maintenance during the shutdown time. Steam and nitrogen are typically used for cleaning out and purge-out operations. A 2-inch nozzle with the valve is provided, so that a steam connection can be made to vent out all the hydrocarbons from the tower. Nitrogen connection can be taken from the utility station, and a hose piping can be run into the tower to ensure there are no hydrocarbons in the tower. Similarly, a utility air connection can be taken from the nearby utility station and connected to the tower to displace nitrogen from the tower.

Typically, all the towers or columns are tall, more than 100 feet, and an operator cannot reach that height routinely. All the parts of the column and nearby equipment such as reboiler and overhead condenser including the column is protected by a relief valve. The relief valve is either provided on the column overhead vapor piping or can be located on the overhead receiver vessel.

The vapor generated from the column goes to the condensers and then to the overhead receiver. Ensuring that the vapors and the condensed liquid are free drained into the overhead receiver is important.

Every column needs several instruments to monitor performance remotely by an operator. The column pressure is the most important parameter that dictates the separation, and any changes from the ideal pressure may impact the product specification. Measuring the pressure drop across the column is important to assess if fouling is happening across the trays. The liquid level at the column bottom helps ensure that the reboiler pumps are getting liquid all the time and the reboiler is not starving off any liquid; otherwise, the heater tubes are overheated. Temperatures across the trays tell a story about the health of the column; for example, if the temperature for one of the top trays is hotter than the other trays at the bottom section, that means a localized problem with the distribution. The tray temperature also ensures that the column produces products to meet the desired specifications.

The columns could have multiple trays for effective distillation. Every tray has a weir plate and a downcomer. Weir plate height determines the liquid level on the tray. The liquid from the previous tray is transported to the tray below using the downcomer pipe. The packed tower has a packed bed. One packed bed tower may have more than one packed beds. The packing material can be structured or random packing. For towers with much smaller diameters, packing is preferred over the tray as a mass transfer media. Figure 2.9 shows the tray and packed bed tower arrangements on a P&ID.

The P&ID is a place where most of the details regarding the column should be shown but not the greater details, such as tray type or weir height or the downcomer pipe size, etc. The P&ID should mention the equipment size, the design detail, the nozzle sizes, the connecting piping, all the required instrumentation, the number of trays or the number of packing beds, any relief protection, details regarding the reboiling and the condensing equipment, and details regarding the pump around pumps and column bottom pumps.

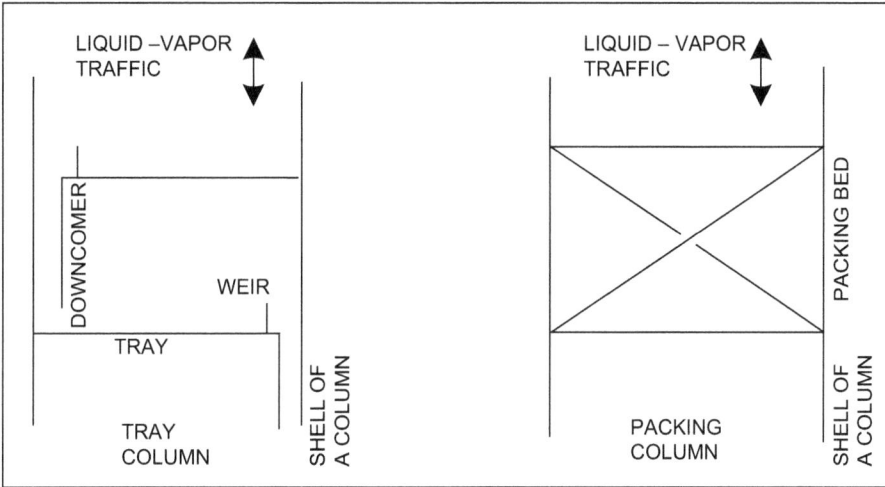

Figure 2.9: Arrangement of a column on P&ID.

Apart from the column trays or packing details, the column also has details of redistribution tray, reflux distributor, feed pipe distributor, reboiler return piping distributor, chimney type of trays, and distributor, which need not be shown in detail on P&IDs; however, a general intent should be shown.

Sections 11.12,11.119, and 11.25 explain a distillation column and the overhead system.

2.11 Turbines

Turbines are used primarily for two purposes. One of the purposes is the expansion of gases in the process, and the other is to make power by letting down high-pressure steam into low-pressure steam.

Traditionally, steam letdown stations are used in the industry to produce medium-pressure and low-pressure steam from high-pressure steam. A significant amount of energy in the form of pressure is wasted in the letdown using a control station and cannot be recovered. On the other hand, steam turbines can rotate the turbine shafts using high-pressure steam, which produces power. Also, medium-pressure and low-pressure steam can be generated, which can later be integrated within the processing plant for further usage.

The turbine needs superheated steam as the saturated steam has some water, which can damage the turbine blades at a very high speed. The boiler produces super-heated steam by overheating the saturated steam from the steam drum.

Since most of the steam usage in the processing plant is saturated, it is beneficial to use a turbine, which also produces saturated steam.

The operation of a turbine is tricky and needs to be studied carefully before starting or shutting down the turbine. As the turbine rotates at a very high speed, any sudden changes in the steam pressures can cause catastrophic failure of the turbine blades. For this reason, the turbine needs to be started or shutdown very slowly. In addition, there are safety controls that trigger a shutdown of the turbine, if the turbine control module sees any abnormal situations such as low flow of steam, low lube oil pressure, high vibration, etc.

Adding a steam trap to remove any possible condensate before the steam enters the turbine is always beneficial. Also, it is always a common design practice to insulate the turbine itself and the steam inlet and outlet piping, so that the heat loss from the steam can be minimized to make maximum power for a given steam rate.

A turbine can also fail and the high-pressure steam can overpressure the downstream piping of the turbine; to take care of these scenarios, safety valves are always installed on the turbine downstream to protect from any overpressure.

Chapter 3
Development of piping on P&IDs

Referring to the PFD, process engineers can connect different pieces of equipment from one page to another. Once all process equipments are laid out on other pages of P&IDs, additional process equipments can be connected by pipelines. This chapter is about developing piping by adding pipe specifications required for the design piping, and several thumb rules used for effectively communicating the piping design information on the P&ID.

3.1 Introduction to piping

Piping is a cylindrical metallic piece, varying in diameter from 1 to 4, 6, 8, and 10–30 inches in diameter. The length of the piping depends on the flow through the pipe. The piping material comes in different materials of construction; the most common being carbon steel and different alloys of carbon steel. Stainless steel can be used for corrosive services, but it is expensive. For high-temperature services, 5 chrome alloy steel can be used. Piping is used everywhere in the industry.

There are natural gas pipelines all over the country and the world. Varieties of liquid, gas, slurries, and utility fluids are transported through the pipelines. All refineries transport liquid hydrocarbon fuels, natural gas, propane, butane, and gas fluids. All utilities in chemical, refinery, and domestic industry also use carbon steel for piping. Utilities include cooling water, steam, nitrogen, and different types of water. Moreover, crude oil is transported through pipelines from one place to another.

Two pieces of equipment that have nozzles are connected by piping to transport fluid from one vessel to another. One piece of piping or more than one piece is either flanged or welded together. Welds are common for all pipelines, and flanges are also common, but they tend to leak.

Tubing is smaller than one inch size. They are used in the industry to transport a low volume of utilities or for instrumentation purposes. Air is supplied to all instruments in all industries using tubing. Tubing is usually field-routed by technicians. The tubing usually connects different parts of the sample station. Different types of analyzers and flowmeter tappings are made from tubing. Instrument air is routed in the plant using stainless steel tubing. Since a tubing is designed for very high pressure, such as in thousands of PSIG, their use is very robust in all smaller volume services.

https://doi.org/10.1515/9781501519864-003

3.2 Significance of piping

In the very old days (1930), companies used to transport crude oil in several barrels using a wood-based transportation method. With the improvement in technology, when the rail came along, and with the invention of oil exploration, rail car transportation became the most common transportation method. This has continued for a long time. Rockefeller figured out in the early days that railcar transportation owners were charging a premium rate for the transportation of oil barrels. Then, Rockefeller came up with a new method of transporting oil, through pipelines. To this day, fluid is transported using pipelines; it is an efficient mode of transportation. Oil transportation through piping became very economical, saving transportation costs and reducing or minimizing environmental emissions that one would have otherwise seen through rail car transportation. Piping not only takes less space, compared to railcars, but is also a highly efficient transportation means. A single pipeline could do the oil transportation job in a couple of hours, which would otherwise need 100s of rail cars.

3.3 Pipeline tagging and information

Each pipeline in the industry has a tag that offers information such as line size, insulation, tracing, fluid code, piping material, etc.. The line tag is like a biodata for pipelines; so the line tag is used to identify the service on P&IDs. The line tag is also helpful in developing the line list, where process and piping engineers input the information regarding the line in the process operation. Once the line list is developed, the information is passed on to develop the isometric drawings. The isometric drawing is then used for the fabrication of different pieces of piping in the process unit.

The line codes are typically applicable for pipes of diameter one inch and more. Note that the pipeline code could differ for customers and site locations. Pipeline codes are commonly specified as 25-10"-ABCD-M-2005-2"-ET. Below is the significance of each item –

- 25 – Unit number
- 10" – line size
- ABCD – pipe specification (shows class and metallurgy)
- 2005 – line number
- M – Material in a pipe (could be process or utility)
- 2" – thickness of insulation
- ET – type of tracing (could be electrical or steam)

3.4 Line size

Line size varies in all chemical or refining plants, depending on the service and flow rate in the pipeline. The service can be liquid, two-phase, vapor or gas, and corrosive. The liquid line is normally sized for 6 to 8 feet per second velocity; the common example is water. The two-phase flow in a pipe is sized for 30 to 35 feet per second; a very common example is flashing condensate. Flashing occurs downstream of a pressure-changing device or instrument, such as a control valve. Hence, minimizing the distance between the flashing control valve outlet and the connecting equipment is always recommended. Gas or vapor has a much lower density than liquid, and takes more volume in the pipe than liquid. Gas pipelines are sized for 100 to 130 feet per second; a common example is steam. On the other hand, a corrosive service is sized for somewhat lower velocities, such as 4 to 5 feet per second; a common example is an amine. The lower velocity for the corrosive service ensures that the corrosion rates are accounted for in the long run of the plant.

Another criterion to size a line is the $\Delta P/100$ feet. The main header, such as cooling tower supply or heat hot oil heater supply, is sized for 0.25 to 0.3 PSI per 100 feet. The sub-headers, such as cooling water from the main header to the individual heat exchanger, are sized for a somewhat higher pressure drop/100 feet of 1 PSI/100 ft. Special lines in the process, such as reboiler piping and pump suction, are always sized conservatively to minimize the pressure drop as much as possible, to achieve an efficient process operation.

3.5 Selection of piping specification and guidelines

Different materials are used in the industry, depending on the service or the fluid handled. The different materials considered are carbon steel, stainless steel, and alloy steel. Among these, carbon steel material is most commonly used for piping and transporting fluids. For corrosive services, stainless steel is used. For high-temperature applications, for example, fired heater return pipe to the column, alloy steel is used. Acids such as HCl corrode all materials except Hastelloy and plastic.

A piping specification document is prepared for all projects. Customers usually have the service index; if it is not available, the process engineer prepares one that is specific for the project, covering all the services used. The piping service index consists of the service to be handled, the code for the actual pipe specification to be used, the metallurgy of the actual piping, the corrosion allowance, any comments or notes, and revisions that go with the service index. An example of such a piping service index is shown in Table 3.1.

Table 3.1: Example of a pipe service index.

Service	Pipe spec (PS)	Pressure rating	Pipe material	Corrosion allowance	Comments	Rev.
High-pressure steam (700 PSIG)	HPS01	600#	Carbon Steel	0.063"	None	A0
Soft water	SW02	150#	Carbon Steel	0.063"	None	A1
Demineralized water	DM02	150#	Stainless Steel	0.063"	None	A1

3.6 Selection of piping class

The class of piping is the rating of the piping. When the rating is lower, the piping can handle lower pressure and temperature. The piping class depends mainly on the pressure and temperature of the fluid inside the pipe. For example, a 150 class piping can only allow a fluid with a temperature of 400 °F at 200 PSIG. Table 3.2 shows the values of maximum the hydrostatic test pressure for different temperatures and pressure classes of the pipe flanges. This is a beneficial information in determining the class of the piping during the markup of the P&IDs.

Table 3.2: Changes in pressure and temperatures with pipe class.

Temperature (°F)	Maximum allowable non-shock pressure (psig)						
	Pressure class, pounds						
	150	300	400	600	900	1,500	2,500
	Maximum hydrostatic test pressure (psig)						
−20 to 100	285	740	990	1,480	2,220	3,705	6,170
200	260	675	900	1,350	2,025	3,375	5,625
300	230	655	875	1,315	1,970	3,280	5,470
400	200	635	845	1,270	1,900	3,170	5,280
500	170	600	800	1,200	1,795	2,995	4,990
600	140	550	730	1,095	1,640	2,735	4,560

3.7 Smart plant 3D

Smart plant P&IDs are also intelligent P&IDs. These P&IDs connect one P&ID to another so that the user can follow the flow of the process. The smart plant P&IDs provide information on the valve; information from the vendor can be embedded into the valve and it can be looked up for reference. The equipment design information, for example, design parameters, can be looked at while looking at the P&IDs. For example, the equipment type of a heat exchanger can be viewed using smart plan

P&IDs. Detailed piping information, such as the pipe size, insulation, pipe specification, etc., can be directly seen from the smart plant P&IDs. A 3D design of the plant can be interfaced with the smart plant P&IDs. Smart plant P&IDs are commonly used across all projects for all processing plants to speed up the design progress in all phases.

3.8 Account for future expansion

Initially, the unit's pipeline is designed for the required capacity of the plant. Often, there are projects to be upgraded or revamped that would either need a new tie-in, a cold work, or a hot work on a pipe. All of these can be avoided by either adding a blind flange or a simple valve, with a note for future connection or expansion. An additional valve or a flange on a piece of equipment or tank is helpful for any future addition or piping connections.

Chapter 4
Control system components

Many components in the instrument system provide measurement, indication, safety protection, and control of some process property using a device or instrument. The components are mainly divided into in-line instruments and offline instruments. Once the piping is connected to the different equipment, a process engineer can start adding details of in-line and offline instruments. In-line instruments are present on the line, for example, a flow meter that has an orifice inserted into a flange and attached to a tubing to measure the pressure drop across the orifice. An offline instrument is not part of the line. A common example of such an instrument is a pressure gauge, which measures the pressure of the pipeline or equipment without altering the fluid flow inside the pipe. Some safety protection instruments, such as safety valves, are provided to protect a piece of equipment or a system.

4.1 Introduction

An in-line control instrument is installed on the pipeline, and offline instruments are not connected to the pipeline.

The control system offers process parameters such as level, pressure, flow, temperature, and composition. These process parameters are required to maintain a product specification or get the required environmental emissions under limits. Below are some examples of control parameters that help in understanding the logic behind the control parameters.

Level control, as the name suggests, controls the level of the drum, or the column, or the vessel. When the level in the drum exceeds the required set point, and if the level is not controlled, the liquid can enter a gaseous system, causing operational and hazardous problems. As an example, liquid control is always installed on the suction knockout drum of a compressor. This control valve always removes any excess liquid from the drum. This control action saves the compressor from being damaged by the liquid hammer.

Pressure control is about controlling the pressure of a column or line. Column pressure is very important, as a change in pressure also changes the temperature profile of the column. A change in the temperature profile of the column affects the product specification. Hence, it is important to maintain the desired pressure in the column to meet the product specifications.

Flow control, basically controls the flow of the line. A common example is a flow control valve installed on a pump discharge. The flow control on the pump discharge makes sure that the pump operates at the desired operating point of the curve. If the

https://doi.org/10.1515/9781501519864-004

flow control valve is not available, the pump operation cannot be controlled, which can cause damage to the mechanical parts of the pump.

Measurement of a process parameter is required for two reasons. The first reason is to compare the measured value against a set value. The set value of a process parameter is set by the panel operator, based on the predetermined design of the process. The measured value from the instrument is compared against the set value. The difference between these two parameters is then automatically fed to the control valve as a set point. The second reason is that the measured value can be used as a parameter to check the process plant's health and determine if a control action needs to be taken. Monitoring of levels, pressures, composition, temperature, flow, etc., are required to analyze the plant's health. The measurement and indication can be of two types. One is a local measurement; an example is a pressure gauge located locally on a piece of pipe or equipment. The second type of measurement is the pressure transmitter, which measures the pressure and transmits the data back to the computer in the control room. The panel operator in the control room can visually see the pressure parameter on the screen. This is also called the board parameter.

4.2 In-line devices

4.2.1 Valve types

A control valve is an instrument device that controls the process parameter. A control valve is also called a final element of the control or physical element available in the field, which controls a process parameter. When the measured value is compared against the set value, the offset value or the set point is given to the control valve as the set point. If the set point is to increase the opening of the control valve, the control valve body physically opens more to allow the fluid to pass, and vice versa.

Control valves come in different sizes and types. The most common control valve type is a globe-valve type. The type of control valve is determined by the type of action needed for the control. For example, if a quick pressure control or action is required, then a quick opening control valve body is selected. For a quick opening valve, 20% of the valve opening provides 60–70% of the flow. The control valve can be an equal percentage opening, meaning if the control valve is 50% open, then it means that 50% of the flow is passed through the control valve.

Control valves can also be classified based on the action and the input. The types of valves are fail-to-open and fail-to-close. Failed-to-open valves are colored green. Green valves are designed to open 100% when the air supply fails. Failed-to-close valves are colored red. Red valves are designed to close 100% when the air supply fails. As an example, the reflux valve to the distillation column is always green colored because the reflux action cools the hot column during an emergency event. Feed to the process unit is always designed as a red valve to cut off feed coming to the unit during an emergency.

The on-off valve is quite different from a control valve. The primary action of the on-off valve is to either completely open or completely close a part of the piping. On-off valves are used mostly in oil and gas and in batch operations. On-off valves are used in oil and gas for safety reasons. For example, during an emergency, if there is a fire in the processing unit, every piece of equipment or section of the piping is provided with an on-off valve, which vents the entire inventory to the flare system. On-off valves are located on the inlet and outlet streams of the equipment, which also close 100%, ensuring that the vessel is isolated from the nearby equipment.

There are different types of on-off valves and the choice depends on the type of on-off action needed. The common type of on-off valve used in the oil and gas industry is the plug-type valve because it reduces pressure drop across the valve when the valve is normally in operation. Butterfly valves are used when a tight shut-off is not needed; they are also cheaper than other types.

On-off valves are linked to the plant's cause-and-effect safety shutdown system. If a cause were to happen, it could close or open multiple on-off valves at the same time as part of the design.

Motor-operated valves (MOV) are also on-off valves, which are not linked to any safety shutdown logic. Typically, MOVs are installed on big pipe sizes, where closing the valves might need more than one person, or the valve could be very remotely located. In such cases, MOVs are more economical and operationally friendly. The MOV has a valve body with a motor attached to it. Control is achieved through the panel. The size of MOV can vary anywhere from 16 inches to 52 inches. A common example where MOVs are used is big cooling water headers of 42-inch pipe size. It needs several operators to close the 42-inch valve, and the valve may not be accessible to operators on the field. Another common example is a 52-inch plug-type valve located on the natural gas transportation line, which could be located on the gas processing platform.

Another type of valve is the automatic recirculation (ARC) valve. These valves are used on the centrifugal pump discharge to protect the pump from overheating, cavitation phenomena, and from running the pump dry. When the pump exceeds the minimum flow required for the pump, the internal mechanism of the ARC valve allows the fluid to be pumped in the forward direction of the process. When the pump reaches a flow that is below the minimum flow required for the pump, the internal mechanism of the ARC valve closes the forward action of pumping and diverts the minimum required flow back to the process, thus making sure that the pump has the required minimum flow always. A common application of the ARC valve is in the tank farm area, where pumps could be sitting idle for days or hours waiting for the pumping action to occur for the barge or rail car. In such cases, the ARC valve, as an alternative to the control valve, could be useful to protect the pump from damage.

Backpressure regulators are somewhat similar to control valves. The stem controls the control valve, which is regulated by the instrument air-operated diaphragm. On the other hand, pressure regulators also have a diaphragm, but a screw manually controls it. For example, suppose the operator needs to adjust a set pressure of the regulator, the operator can turn the screw to increase or decrease the pressure on the diaphragm of the pressure regulator, affecting its set pressure. The primary purpose of the backpressure regulator is to manually maintain constant pressure on the pipe or equipment.

Table 4.1 summarizes the P&ID symbology for different in-line control system components used in the industry.

Table 4.1: Different in-line control system components and symbols.

Valve type	P&ID symbol
Diaphragm-type control valve	
Diaphragm-type control valve with a hand wheel	
On-off valve	
MOV	
ARC	
Backpressure regulator	
Pressure-reducing manual valve	

4.2.2 Flow measurement and restriction

There are different instruments available to measure flow through a pipe. The type of instrument to be used depends on various factors such as flow accuracy, pressure drop allowed, range, mechanism of measurement, etc.

An annubar flowmeter involves an instrument inserted into the pipe. It has tubing on the front and back side. The front side measures the high side of the impact pressure, and the back side of the instrument tubing measures the low side of the pressure. The differential between the high and low side impact pressures is the differential pressure measurement. The differential pressure can be correlated to the flow through the pipe. The annubar takes very little space of the piping cross-sectional area and the pressure drop across annubar is very small. Annubar element is also commonly fitted with the temperature thermocouple, and changes in fluid temperature can be integrated into the fluid flow measurement. Annubars are widely used in the industry for gas, liquid, and oil applications.

The ultrasonic flowmeter works on the principle of measuring the travel time of the ultrasound between two sensors. Different sensors are installed around the pipe. A constant ultrasound is sent and received through these sensors. As fluid passes through the meter, the travel time of the ultrasound is affected. For example, the travel time for the fluid with a high velocity is lower than that for the fluid with a lower velocity. The ultrasonic flow meter can be installed directly on the pipe as a clamp-on device, without interfering with the flow of fluid inside the pipe. These meters are generally expensive and can only be used for fluids with temperatures below 200 °C. Ultrasonic flow meters can be used in oil and gas, water, chemicals, and pharmaceutical applications.

A vortex flow meter works on the principle of creating vortices inside the meter using a bluff object. The bluff body creates a vortex-like shape as the fluid passes through the meter. A device installed to the meter can detect the distance between two vortices, and the flow is measured. At low velocities, lower quantities of vortexes are formed, meaning the flow is on the lower side of the pipe. If the velocity in the pipe is too small, the meter size can be reduced to increase the velocity through the meter, and it does not affect the accuracy of the measurement. Temperature measurement and the vortex detection device make the vortex meter useful in various industrial applications where temperatures vary. The common applications are steam, gas, and liquid applications. The drawbacks of a vortex meter are that it does not have good vibration resistance and cannot handle dirty fluids.

Flow orifice-type meters are commonly used industrywide and are a reliable way of measuring flow through the pipes. It involves measuring the pressure drop across the orifice. The measured pressure drop is correlated to the flow rate using eq. (4.1).

$$(\Delta p) \, \alpha \, (Flow)^2 \cdots \tag{4.1}$$

Orifice-type meters are commonly used in the industry. There are two tubings, one on each side of the orifice. These tubings are connected to the main pipe. As the fluid moves through the orifice, a pressure drop is created across the orifice. The fluid also moves through these tubes, creating a pressure differential across the diaphragm. The change in pressure across the diaphragm is calibrated, and the measured flow is recorded. The drawback of an orifice-type meter is that the meter takes a certain pressure drop across the orifice, and needs a single phase and homogeneous mixture. It also needs an axial velocity vector, downstream of the meters, for better accuracy.

Unlike other meters, the primary job of restriction flow office is to restrict a flow to the desired value. Restriction flow orifice involves stainless steel orifice plate sandwiched between a pair of flanges. The size of the orifice depends on the pressure drop to be taken from the system. A common application is in the pump-minimum circulation lines.

The operation of the Coriolis flow meter depends on the principle of fluid mechanics. Fluid is forced to accelerate as it moves through the tubes, which vibrates as it travels, and decelerating fluid moves away from the peak amplitude. The density of the fluid is a function of the sine wave; higher the frequency, lower is the density, and lower the frequency, higher is the density of the fluid. The Coriolis meter is widely used, but accurate measurements are necessary for safety and accountability. These meters are commonly used in custody-meter applications. These are also used in applications where the fluid composition is unknown and could change.

The magnetic meter measures an electrical signal, which can be related to the fluid flow rate inside the pipe, across the magnets. As the electrical signal is proportional to the flow rate of the fluid, it is easier to measure the flow rate of the fluid. The electrical signals can be calibrated with the flow rate. The meters have no pressure drop as there are no moving parts. Also, the meter and the whole assembly match the upstream and downstream pipe sizes fairly well. Thus, upstream and downstream straight pipe requirement is not applicable for this meter. The main drawback of the magnetic meter is that the fluid should have some conductivity; otherwise it cannot be detected by this meter. The flow rate of hydrocarbons or organic fluids cannot be measured through the magnetic meter as these fluids have no conductivity; common applications are water and ionic services.

A turbine flowmeter works on the simple principle of rotation of a turbine. This meter has a simple tube, with a turbine inserted into the tube. As the fluid moves, the turbine rotates and generates pulses. These pulses are picked up by a device installed along with the meter. The turbine generates several pulses. The greater the number of pulses, higher is the flow rate. Turbine flowmeters are very accurate, but they can only be used in very clean applications, such as clean water. They cannot be used for high-viscosity applications such as vacuum gas oil. In applications with a lot of impurities and dissolved solids, these can damage the turbine element. The turbine flowmeter requires an upstream straight pipe length of 10 diameters and a downstream straight pipe length of 5 diameters.

The venturi flowmeter has a venturi body, where the flow is restricted at the throat, increasing the fluid's velocity. So, the pressure drops across the unrestricted flow areas and the restricted flow areas can be measured to determine the flow rate through the pipe. The throat area of the venturi meter can be designed to meet the required accuracy of flow measurement. Venturi meters are mostly used in applications where the lowest pressure drop through the meter is required. Venturi meters take up a lot of space and cannot be used where the space is limited. It also requires upstream and downstream straight pipe length requirements, which may not be possible in all situations, and venturi meter maintenance is not easy. A common example is the sulfur recovery unit, where the lowest pressure drop is required, as the process system operates at only 6 to 8 PSIG.

Rotameters are used to measure the flow rate of gas or liquid; the measurement principle is the volumetric rate of incoming gas or liquid using a rotameter device. If there is no flow, the float inside the rotameter stays at the bottom, indicating no flow. As the flow increases, the rotameter moves in an upward direction until it attains equilibrium, which is the fluid's flow rate inside the pipe. Some of the drawbacks of rotameter are that the rotameter needs to be mounted vertically, the flow measurement accuracy is not so great, and the applications are only limited to small-bore pipe sizes. A common example is measuring the chemical injection dosage. Table 4.2 shows the P&ID representation of different types of flow measurement and restriction instruments.

Table 4.2: P&ID symbology for different flow instruments.

Type of flow measurement and restriction	P&ID symbol
Annubar flowmeter	
Ultrasonic flowmeter	
Vortex flowmeter	
Flow orifice	
Restriction flow orifice	
Coriolis meter	
Magnetic flow meter	
Turbine flow meter	

Table 4.2 (continued)

Type of flow measurement and restriction	P&ID symbol
Venturi flow meter	
Rotameter	

4.2.3 Relief

Relief valves, also commonly known as safety valves, are installed in refineries, oil and gas plants, and chemical plants to protect a piece of piping or equipment. These valves are essential for processing plants as they protect the equipment or piping from overpressure. Generally, vessels and piping are designed for a certain pressure, which makes economic sense, and a lower design pressure is preferred from the cost perspective. To keep the design pressure of the system lower, a common practice is to use a relief valve, and set it at the design pressure of the system.

There are several types of relief valves. The choice of the type of relief valve to be used depends on the type of service and the backpressure generated by the system. Conventional relief valves are very common but are limited to only 10% backpressure on the outlet system. Bellow valves can handle 25% to 40% backpressure in the outlet system and are very good alternatives to conventional valves. Bellow valves are in high demand as the allowable backpressures are higher. Pilot-type relief valves are used when the operating pressures are very close to the set pressures. The drawback of the pilot relief valve is that it can be used in clean service as the pilot tubing could choke if any foreign material is received from the process fluid. Sections 11.4 and 11.5 show the overall system of the relief system, with piping and valve connections.

Rupture discs are similar to relief valves. The main difference between the ruptured disc and the relief valve is the relief mechanism. The relief valve has an internal spring mechanism, calibrated to a certain set pressure. This spring is activated when the pressure inside the relief system exceeds the assigned set pressure. On the other side, the rupture disc works on the disc principle. The disc is made of a metallic component, compatible with the relief fluid or gas. The disc thickness is designed to withstand the set pressure of the relief system. When the system pressure exceeds the set pressure, the disc tends to break, allowing the process fluid to relieve into the relief header. Ruptured discs are used where the rapid pressure buildup is seen in the system and where a corrosive fluid could damage a relief safety valve. A combination of a ruptured disc and a safety valve is also commonly practiced in the industry to avoid the fluid directly touching the internal parts of the safety valve. Rupture pin valves are very similar to rupture discs. The only difference is that the rupture pin valve has a metallic pin, designed to withstand a set pressure of the relief system. As the pressure in the relief system exceeds the set pressure of the valve, the pin inside the rupture pin valve buckles, allowing a rapid depressurization of the relief fluid.

Below is an example that shows how the design pressure of the system is assigned in conjunction with the set pressure and the type of the relief valve. The high-pressure side of the exchanger is designed for 1,000 PSIG, and 1,000 PSIG value is designed based on the process-side reactor design pressure and the pump deadhead conditions. The exchanger is cooled by a cooling water supply. The cooling water-side system is designed for 150 PSIG, based on the deadhead conditions of the cooling water pumps. If the cooling water-side design pressure is increased to at least 1,000 PSIG, the exchanger cost increases, as the cooling water-side, which is located on the tube-side, will have to be designed for 1,000 PSIG. On the flip side, the cooling water-side can be designed for 150 PSIG, and a relief valve can be installed to relieve any overpressure caused by the high-pressure side, thus reducing the cost of the cooling water-side of the exchanger. However, it is important to note that the process fluid at 1,000 PSIG will be relieved very rapidly into the cooling water-side during a tube rupture relief scenario. Suppose the over-pressuring fluids are not relieved from the cooling water-side during this overpressure event, the cooling water of the exchanger could be subjected to a catastrophic failure, which could relieve hazardous chemicals or gases into the atmosphere. A rupture pin valve is most suitable in this application to rapidly respond to the tube rupture scenario.

Table 4.3 summarizes the P&ID symbology for different types of relief valves used in the industry.

Table 4.3: P&ID representation of different types of relief valves.

Type of relief valve	P&ID symbol
Spring operated pressure safety valve	SET @ XX PSIG
Pilot safety valve	SET @ XX PSIG

Table 4.3 (continued)

Type of relief valve	P&ID symbol
Rupture disc (RD)	
Rupture pin (RP)	
Pressure and vacuum relief valve	

4.3 Offline instruments

Offline instruments measure process parameters that are helpful in assessing the health of the process unit operation. Offline instruments are categorized as mentioned below.

4.3.1 Gauges and switches

A pressure gauge is a simple instrument that measures the pressure of the piping or the vessel. The mechanism involves a bourdon tube and some mechanical moving parts inside the pressure gauge housing. If the pressure is below the atmospheric or if it is vacuum, the bourdon tube shrinks, and the pointer moves to the left, indicating vacuum in the system. On the other side, as the pressure increases in the system, the bourdon tube expands, and the pointer moves to the right side of the pressure gauge monitor. A good pressure gauge has a range that shows the pointer in the middle portion of the gauge monitor for the system's operating pressure. Pressure gauges vary in pressure ranges from 0 to 20,000 PSIG.

Sometimes, the process fluid may be impure, which directly interacts with the internal components of the pressure gauge, and may cause severe damage to the instrument. A diaphragm, coupled with the pressure gauge instrument, is typically used to avoid this damage. The diaphragm is made of stainless steel, Hastelloy, or of Inconel material. As the pressure in the system changes, it changes the position of the diaphragm, which in turn changes the position of the bourdon tube.

The process fluid or media has solids that can clog or damage the equipment, such as pumps. It is important to capture such solids before they cause any damage to the downstream equipment. To capture them, simple devices such as filters or strainers are installed. Over time, the solids clog up the strainers and filters, increasing the pressure drop across. So, if the pressure drop across the filter is left unmonitored, the solid can escape and can cause damage to the downstream equipment. Pressure drop across the filter or the strainer is measured to check the health or the condition of the filter or strainer element. Differential pressure (DP) gauges are installed across the filter or strainer to measure the pressure drop. The DP gauge has a high-pressure side and a low-pressure side. The pressure on both sides of the filter is first applied to the diaphragm. The change in diaphragm position is mechanically captured in the DP gauge housing. The pointer on the DP page moves to the right as the filter or strainer gets blocked with material.

A pressure switch is a device that indicates that the pressure in the system has exceeded the desired pressure or the set pressure. Pressure switches are useful to indicate the system's abnormality, and alarm the process operations before a catastrophic failure event. The pressure switch involves a diaphragm, a spring, and a pin. As the pressure in the system increases, it pushes the diaphragm, which in turn creates tension in the spring and, as a result, the spring moves. This moving action of the spring will cause the pressure switch pin to move. The movement of the pin is captured electronically and indicated on the control panel. A similar functionality could be explained for the temperature switch.

Table 4.4 shows the P&ID representation of different gauges and switches.

Table 4.4: P&ID symbology of different gauges and switches.

Gauges and switch type	Function	P&ID symbol
Pressure gauge	Measure and indicate the pressure	
Differential pressure gauge	Measure and indicate diffrential pressure across a strainer or filter	

Table 4.4 (continued)

Gauges and switch type	Function	P&ID symbol
Temperature gauge	Measure and indicate temperaure	TG XXX ⊠ 1/2"
Pressure switch	Indicate pressure	PS XXX
Temperature switch	Indicate temperature	TS XXX

4.3.2 Transmitters

Thermocouple-type temperature transmitters are very common in the industry and are very cost-effective compared to other temperature transmitters, such as resistance temperature detection (RTD). The primary function of the temperature transmitter is to indicate a temperature reading on the control panel. The working principle of the temperature transmitter involves a metal rod inserted into the process piping or vessel. The metallic component of the rod has two types of metals. One side of the thermocouple is at a fixed temperature, and the other side sees the process fluid temperature. As the process fluid temperature changes, a differential voltage is created across the two types of metals. The differential voltage created can be used to estimate the process temperature inside the piping or equipment. Most used metals are base metals, limited to lower process temperatures. Noble metals are commonly used for high-temperature applications.

Current-to-pneumatic (ITP) converters are used widely in the industry. They convert the electrical signal from the control board to the pneumatic signal required for the control valve function. They are required because the control valve cannot read the electrical signal, and only functions when pneumatic signals such as air pressure are given to it.

About 4 to 20 milliamps of current is supplied from the control panel, based on the desired set point. The ITP converter converts this current signal into an equivalent amount of instrument air pressure, which is between 3 and 15 PSIG. Based on the pressure signal received, the control valve changes the stem position and does the control function, as desired.

Pressure transmitters are used to measure and indicate the pressure of the system or the pipeline remotely inside the control room. A differential pressure transmitter is used to measure differential parameters across a system; they are indicated remotely inside the control room. As the pressure and differential pressure transmitters are expensive compared to their parent gauges, the transmitter should be carefully assigned to the processing plant on an as-needed basis. For example, monitoring the pressure differential across the strainer located at the pump's suction may not be as critical as the pressure differential across the reactor bed. Another example is when the pressure in a pipeline transferring fluid from one unit to another unit may not be that critical, but the pressure inside a distillation column is a critical parameter to watch for.

The pressure or differential pressure transmitter works on a simple component such as a diaphragm and resistor. As the pressure inside the system changes, it changes the position of the diaphragm. The change in position of the diaphragm is recorded by the register. The register's position change can be calibrated to the required pressure.

The transmitter produces the process variable signal. The controller looks at the difference between the process variable and the set point. The controller decides what action the actuator should take. An actuator is a valve that actuates based on the signal.

Table 4.5 shows the P&ID representation of different transmitters.

Table 4.5: P&ID symbology of different transmitters.

Transmitters	P&ID symbol
Pressure transmitter	
Differential pressure transmitter across a strainer	
Temperature transmitter	

4.3.3 Level

Level instruments measure the level of the equipment. The differential pressure (DP)-type level instrument is calibrated from 4 to 20 mA. The following example shows how the level instrument functions. 0 to 150 inches of water level in the equipment is calibrated using a 4–20 mA current.

150 inch water level, specific gravity $(SG) = 1, 24$ mA $= 150 \times 1 = 150$ inch of water
0 inch water level, $SG = 1, 4$ mA $= 0 \times 1 = 0$ inch of water

Table 4.6 shows the P&ID representation of level transmitters.

Table 4.6: P&ID symbology of different level transmitters.

Type of level transmitter	P&ID symbol
DP level	LC DP
Magnetic	LC MAG
Radar	LC RAD
Ultrasonic	LC ULT

4.3.4 Analyzers

Analyzers are used to measure a certain process parameter, as shown in Table 4.7.

Table 4.7: Different types of commonly used analyzers and their function.

Type of analyzer	Function
pH	Measure the pH of the solution
Conductivity	Measure the conductivity of the fluid
Moisture	Measure the moisture content in the fluid
Cl	Measure chlorine concentrations
HC	Measure hydrocarbon concentrations
H_2S	Measure hydrogen sulfide gas concentrations
NOx	Measure concentrations of NO and NO_2 gases

Chapter 5
Process controllers, alarms, and safety instruments

Process controllers control a certain parameter required for the process. Process alarms alert the process operators about abnormal situations in the unit operation. Safety instruments protect the plant from unsafe situations or scenarios. These instruments are critical to running a processing plant safely. All such instruments are automated and do not require a human interface, once designed.

5.1 Process controller types

The process controller maintains a certain process parameter, based on the target value provided by the process operator. One controller can provide a set point to another controller in a cascade control scheme. Table 5.1 shows some process controller types and their function.

Table 5.1: Process controller types and their function.

Type of a controller	The function of the controller	Representation on a P&ID
Flow	Controlling flow	FC
Temperature	Controlling the temperature	TC
Pressure	Controlling the pressure	PC
Differential pressure	Controlling the differential pressure	DPC
Level	Controlling the level	LC
Reflux ratio	Controlling the reflux ratio	RY
Composition	Controlling the composition	CC

https://doi.org/10.1515/9781501519864-005

5.2 Cascade process controls

Cascade control involves two controllers; one acting as a master and the other acting as a slave. Even though there are two controllers, the controller's objective is one parameter. The master controller provides the necessary set point to the slave controller. Based on the input from the master controller, the slave controller adjusts its set point and provides a final control action to the control valve. Such an example is explained using Figure 5.1. The primary target of the controller is to control the level in the tank. A flow control valve controls the tank level, which takes the set point from a flow controller. The flow controller receives the set point (SP) from the level controller of the tank.

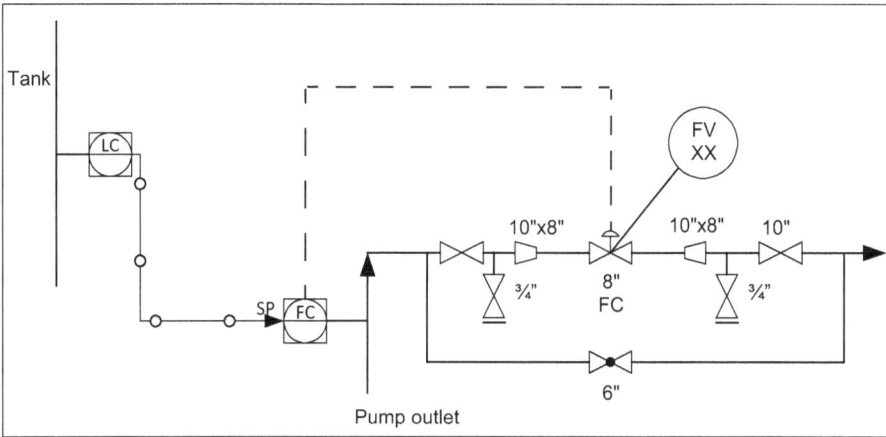

Figure 5.1: Example of a master and slave controller.

5.3 Process alarms

Process alarms alert the process operator of any abnormal situations in the processing plant. This alert is important so the operator can take necessary action before the plant is completely shutdown or a catastrophic failure occurs. There are different types of process alarms, depending on the process unit operation. Typically, the board instruments such as pressure, flow, and temperature can program a certain process alarm. The alarm set point needs to be below that of the process's major significant process parameter. For example, the level in the knockout compressor drum should not exceed 60%, as an increase in the level can carry the liquid along with the gas into the compressor, damaging the compressor. Moreover, the operator needs enough time to act and figure out the potential root cause for certain level increases and determine the possible solutions to lower the level to a safe level point immediately. Table 5.2 shows some process alarms, and go-by example set points used in the industry.

Table 5.2: Examples of process alarms and set points.

Alarm type	High alarm set point	High-high alarm set point	Low alarm set point	Low-low alarm set point	Significance
Level	60%	80%	30%	20%	High and low levels can damage the downstream and upstream equipment
Temperature	250 °F	300 °F	200 °F	180 °F	High and low temperatures can cause operational problems
Pressure	300 psig	280 psig	150 psig	130 psig	High and low pressures can cause operational issues with the supply of a gas

5.4 Cause-and-effect table

Each processing plant in the unit has a certain level of hazards, based on some causes. Every cause affects the plant, and is documented in the cause-and-effect table. The

Table 5.3: Example of cause and effect.

Unit/area		Area # 1, Unit A		
Interlock No.		1	2	3
Tag no. XX		ES-XXXX	FAL-XXXX	FALL-XXXX
Tag no. XX-XX	Cause / Effect	Emergency shut-down unit A	Flame failure in main heater	Low natural gas flow
XV-XXXX	Clean gas to main burner	C	C	C
PV-XXXX	Natural gas to main burner	C	C	C

cause-and-effect table has more than one column and more than one row. The row on the top shows different causes in the process area. The column toward the left shows different effects that the process engineers identified during the HAZOP operational studies. Some customers prefer to show the cause-and-effect table as part of the P&ID, and the associated interlock number is also shown on the cause-and-effect table. This is also reflected in the actual interlock, next to the process equipment or piping. A typical example of a cause-and-effect table is shown in Table 5.3.

5.5 Safety interlocks

Safety interlock work based on the cause-and-effect table. A specified process parameter triggers the necessary shutdown or action in the process plan, as shown in Figure 5.2 and Table 5.4. In the example shown in Figure 5.2 and in Table 5.4, if the pressure readings by PI-15 are higher than 45 psig or the pressure readings by PI-16 are lower than 3 psig, it shuts down the shutdown valves (SDV), SDV-1 and SDV-2, and SDV-3 opens to protect the system from going into the unsafe scenario.

Table 5.4: Safety interlock table.

	Parameter, PI-15	Parameter, PI-16
Value	45 psig	3 psig
Process element	Process action	
SDV-1	Close	Close
SDV-2	Close	Close
SDV-3	Open	Open

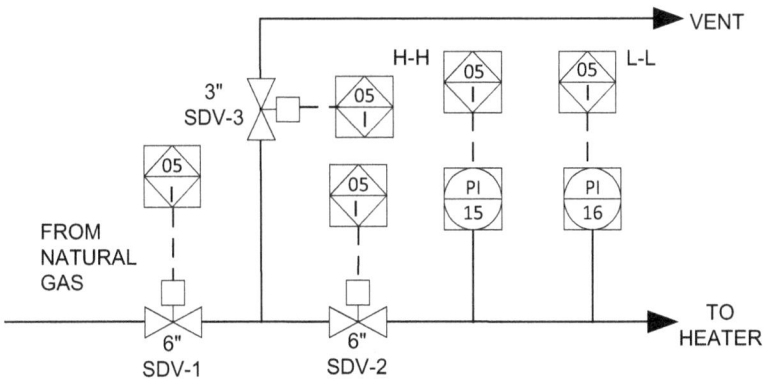

Figure 5.2: Safety interlock P&ID.

Chapter 6
Utilities

Once the equipment are connected through piping and sufficient instrument details are added to the P&IDs, it is time to start adding the details of utilities for the process. As a part of the process design, all utilities are added and shown on the P&IDs. The process engineer can use the PFD as a starting point and create multiple drawings with different utilities and utility distributions, add details, etc. Also, the process engineer can now connect different equipment to the required utility, as required by the design. This section describes the commonly used utilities in process plants, how they are isolated from the process piping or equipment, and the details of utility stations and distribution.

6.1 Introduction to utilities

Utilities are simple process commodities. They can serve as a cooling or heating media that are required for operations. Also, apart from heating and cooling, utilities are used for pressure testing, cleaning, and decontaminating the process vessel or piping to maintain a certain pressure and temperature in the system. Utilities are critical to any processing plant as they are required all the time, irrespective of whether the plant is functioning or not. For example, if the plant is completely shutdown, the engineers would still need steam, nitrogen, and instrument air to decontaminate the hazardous piping or vessel.

Various utilities are designed for specific plant needs. A certain plant type may not need chilled water or a hot oil system. Steam or hot oil is used for heating operations, depending on the process temperature requirement. The distillation column overhead system typically needs a condenser, followed by a trim cooler. The condenser can be an air-cooled or water-cooled exchanger. Column reboiling and any fluid preheating are carried out by steam or hot oil. Cooling water is used predominantly in all industries for cooling services, for example, cooling the product streams going to the tank farm. Low-pressure nitrogen is used for purging out a vessel or blanketing a tank. Instrument air is used predominantly for instrument operation. Fuel gas and fuel oil are used to provide heat to the fired heater. Different types of water are used for either cleaning or operation, depending on the presence of solids and ion concentration in the water. For example, clarified water is used for cleaning the process equipment. The boiler feed water, which is generated from the deaerator, is used to produce steam in the boiler, as the boiler feed water does not have these solids or water hardness.

https://doi.org/10.1515/9781501519864-006

6.2 Different utility systems

6.2.1 Steam and condensate

Different levels of steam and condensate are available in the processing plant. High-pressure steam, available at 600 PSIG, is used for high-temperature applications such as 500 °F. The condensate produced from the 600 PSIG steam is called high-pressure condensate. Medium-pressure steam, available at 250 to 300 PSIG, is used for medium-temperature applications such as 400 to 450 °F. Low-pressure steam, available at 50 PSIG, is used for low-temperature applications such as 300 °F. The resulting condensate from the low-pressure steam is called low-pressure condensate. The high-pressure condensate is flashed in a medium-pressure steam drum, which produces medium-pressure steam. The resulting condensate from the medium-pressure steam drum is further flashed into the low-pressure flash drum. That low-pressure steam drum produces low-pressure steam, which is further tied back to the low-pressure header. The resulting condensate from the low-pressure steam drum is additionally flashed into the atmospheric flash drum. The flashed steam in the atmospheric flash drum is either condensed further or returned to the drum, using a condenser. Any non-flashed condensate in the atmospheric flash drum is pumped to the deaerator drum for further condensate recovery.

Typically, the boiler produces high-pressure steam at 600 PSIG, with at least some degree of super-heat. Super-heat is required as the steam is transported from the boiler to the user; some condensation is expected because of heat loss. Depending on the medium and low-pressure steam requirements, one may have to letdown the high-pressure steam into other levels as required. When the letdown is completed, the steam at a lower level is at a very high temperature, and it may be necessary to bring the temperature down to near-saturation level temperature. This may be done using a de-superheater. The boiler feed water or condensate water can be used to de-superheat the steam. Section 11.2 shows an example of such a de-superheating system.

6.2.2 Low-pressure nitrogen

For an atmospheric tank, when the pump transfers liquid, the tank inhales oxygen from the environment. The oxygen can be hazardous to the chemical inside the tank. When the liquid is being filled into the tank, the liquid constantly displaces the gas in the vapor space, and evaporation of light hydrocarbons to the atmosphere is possible. To avoid this, nitrogen is typically added to the tank in a controlled manner. The arrangement of the nitrogen blanketing system is explained in detail in Section 11.23. The double check valve ensures that the process gas or liquid from the tank does not backflow into the nitrogen utility system, which can be a potential problem for other systems in the process plant.

Split range control is often used for most pressure vessels and columns. The pressure inside the column or the vessel is the most critical parameter that dictates the product specification and composition quality. Flare venting and adding pressure using nitrogen are used to maintain a constant pressure inside the column or vessel. When the pressure in the column or the vessel is lower than the set pressure, the nitrogen valve opens to make up the balance pressure. If the pressure in the column or vessel reaches the set pressure of the system, the flare valve will open and vent the excess pressure to the flare. The basic mechanism is that the controller takes control action to the 50: 50 level. This control mechanism is also known as split range control.

6.2.3 Instrument air

Instrumental air is produced from the atmospheric air. The atmospheric air is first processed using a molecular sieve bed to remove any moisture. Air is then further compressed and cooled. The cold air is further sent through a knockout drum, where further moisture is removed. The clean instrument air is further distributed to all the units in the processing plant. It is important for the units to get around 80 to 90 PSIG pressure at their battery limit to operate the instruments safely. From the battery limit, typically, 2- to 6-inch pipe is routed within the unit. The higher the instrument air demand, the bigger is the pipe size.

Additionally, 2-inch sub-headers are routed to the nearby instruments. From sub-headers, the instrument's tubings are field-routed to the associated instruments. Each instrument sub-header can have several instrument connections, predetermined by customer guidelines. Instrument air tubing is not typically shown on P&IDs; these are

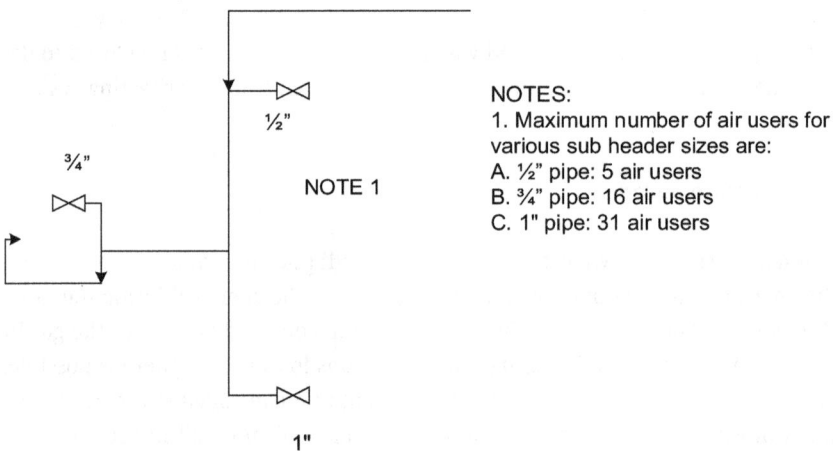

Figure 6.1: Instrument air P&ID representation and guidelines.

field-routed by contractors, based on guidelines. The P&ID representation and guidelines for instrument air usage in a sub-header are shown in Figure 6.1.

6.2.4 Plant air

Plant air comes from the instrument air skid. Drying of air is typically recommended to produce instrument-air quality. For plant air, drying of air is not required, as the application of plant air is not specific, compared to instrumentation applications. Most plants use plant air to drive pumps and other usages in utility stations. During the startup or commissioning activities, initially, nitrogen is used to remove any hydrocarbons or hazardous chemicals from the equipment or piping, followed by low-pressure steam. For an operator or a contractor to enter the vessel or piping, the system needs to have sufficient air or oxygen for breathing. For this purpose, plant air is commonly used.

6.2.5 Cooling water

Cooling water is used for cooling process fluids and gases in the process plant. The cooling water is usually placed on the tube-side of the shell and tube exchanger. Butterfly valves are commonly used in the cooling water service, as most valves are always fully open, and a minimum pressure drop in the headers is expected. The main headers of the cooling waters are sized for 0.5 PSI per 100 feet. For most applications and exchangers, there is no control for the cooling water temperatures. For temperature-critical services or applications, temperature control valves are often used. As the cooling water has some minerals, cooling water-side exchanges are always designed to not exceed 140 °F. Otherwise, the high temperatures can deposit minerals on the tube, causing heat transfer problems. A typical cooling water system is a closed-loop system. The system involves a cooling tower, cooling tower pumps, a strainer in the pump's suction, network piping, and exchangers. Cooling water pumps are always sized to get at least the minimum required cooling water to all users in the system. The cooling water can supply water to more than one unit from a single cooling tower. Depending on the elevations of the exchanges and the grade level of the unit, the units may receive cooling water, and few may starve from the required cooling water. This is a common problem in the industry. To avoid this problem, pressure gauges are installed on the inlet and outlet side of the tube-side of the exchanger. Based on the pressure drop vs. flow curve and the measured pressure drop across the exchanger (using the pressure gauges installed), one can estimate the required flow rate from the curve. The operator can adjust the valve to get the required flow rate, based on the pressure drop. Often, industries use restriction orifices to restrict a certain amount of flow for exchangers located at grade, compared to those in the structure. An example of a cooling water network is shown in Section 11.33.

A very small relief valve, designed for thermal expansion, is installed on the cooling water side. This safety valve accounts for any thermal expansion of cooling water, inside the tube-side of the exchanger.

6.2.6 Clarified water

River water pumps pump river water to clarifiers where total suspended solid particles are removed, and clarified water is produced. The clarified water is then stored in large clarifier water tanks. Then, several sets of clarified water pumps transfer clarified water to the clarified water network. Clarified water is used for utility stations as the makeup for the cooling tower, general cleaning, and as a makeup for the fire water system and tank.

6.2.7 Soft water and boiler feed water

The soft water treatment unit receives filtered water from the anthracite filters. The cation ion exchanger, also called a water softener, removes the calcium, magnesium, and other ions from the filtered water and replaces them with the sodium ions. These heavy ions, such as calcium (Ca^{+2}) and magnesium (Mg^{+2}), if left in water, can cause deposition on the boiler tubes, decreasing their thermal efficiency. Brine water, a mixture of water and sodium chloride, is used as a regeneration media for the water softener. In normal operation, the sodium ions (Na^+) available on the polymeric bed are exchanged with the calcium iron from the water. In the regeneration mode, the brine water with sodium ions replaces the calcium ion on the softener bed. These reactions are explained in eqs. (6.1) to (6.4). The conductivity of the soft water is monitored and this determines when the softener needs to be regenerated.

Brine is pumped using the brine water pumps. Filtered water, or sometimes soft water, is used as motive fluid for the eductor, and brine water is consumed into the eductor and used for the regeneration of the softener.

Ion exchange occurs over a resin bed, represented by 'R' in eqs. (6.1) and (6.2).

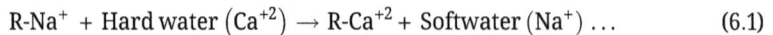

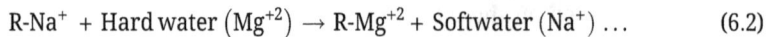

$$R\text{-}Na^+ + \text{Hard water}\left(Ca^{+2}\right) \rightarrow R\text{-}Ca^{+2} + \text{Softwater}\left(Na^+\right)\dots \tag{6.1}$$

$$R\text{-}Na^+ + \text{Hard water}\left(Mg^{+2}\right) \rightarrow R\text{-}Mg^{+2} + \text{Softwater}\left(Na^+\right)\dots \tag{6.2}$$

The regeneration reactions over the softener bed are represented by eqs. (6.3) and (6.4).

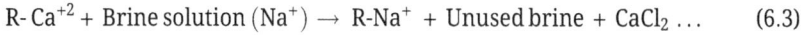

$$\text{R-}Ca^{+2} + \text{Brine solution } (Na^+) \rightarrow \text{R-}Na^+ + \text{Unused brine} + CaCl_2 \dots \qquad (6.3)$$

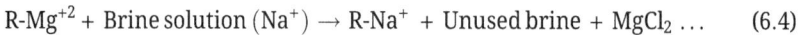

$$\text{R-}Mg^{+2} + \text{Brine solution } (Na^+) \rightarrow \text{R-}Na^+ + \text{Unused brine} + MgCl_2 \dots \qquad (6.4)$$

The soft water produced is stored in the soft water tanks. Depending on the supply and demand, a set of pumps deliver the soft water to the deaerators. The function of deaerators is to remove the dissolved oxygen present in the soft water. If left untreated, dissolved oxygen can cause serious corrosion on the boiler tubes. For the deaerator's operation, soft water is added to the drum, and 25 to 30 PSIG steam is distributed inside the vessel. The oxygen is removed from the top of the vessel through packing. The water produced at the bottom is pumped through high-horsepower feed water pumps to the boiler supply header. As the drum's water is saturated, it is necessary to add the required elevation for the boiler feed water pumps. Based on the design practice, 25 to 30 feet of elevation is required to maintain a net positive suction head for the pump. An overflow line is often designed to remove the excess water flow.

6.2.8 Demineralized water

Similar to the water softener, demineralized water is produced using a combination of cation and anion exchangers. The cation exchanger is regenerated using dilute HCl or H_2SO_4. The anion exchanger is regenerated using dilute sodium hydroxide (NaOH). The regenerated waste is chemically treated in the neutralization system, where acid or caustic is added to neutralize the wastewater to a pH value of 7. The demineralized water is highly pure with no minerals. The demineralized water is stored in a stainless steel storage tank and pumped to the users through a set of centrifugal pumps. Stainless steel metallurgy is important as the demineralized water is very corrosive. The demineralized water is used for applications such as in pharmaceuticals and semiconductor industries.

Water from filters is typically passed through a cation bed, followed by an anion bed. A cation bed consists of a column packed with cation resin. An anion bed consists of a column packed with anion resin. The anion resin can either be a strong base anion (SBA), a weak base anion (WBA), or a mixture of the two. The WBA is used to remove weak organic acids, and the SBA is used to remove strong anions. Once the breakthrough of cations and anions is reached, the cation and anion beds must be regenerated. Based on the type of resin in the bed, regeneration using H_2SO_4 or NaOH is required and is usually specified by the resin manufacturer. Equations (6.5) to (6.8) show some examples of reactions on cation and anion resin beds.

Ion exchange

$$\text{R-}H^{+1} + \text{Cation } (Ca^{+2}) \rightarrow \text{R-}Ca^{+2} + \text{Softwater } (H^{+1}) \dots \qquad (6.5)$$

$$R\text{-}OH^{-1} + Anion\ (Cl^-) \rightarrow R\text{-}Cl^{-1} + Softwater\ (OH^{-1})\ldots \qquad (6.6)$$

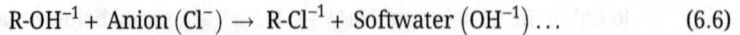

Regeneration

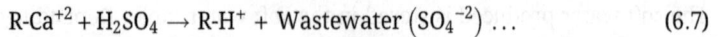

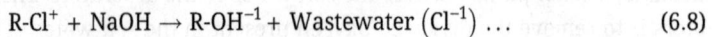

$$R\text{-}Ca^{+2} + H_2SO_4 \rightarrow R\text{-}H^+ + Wastewater\ (SO_4{}^{-2})\ldots \qquad (6.7)$$

$$R\text{-}Cl^+ + NaOH \rightarrow R\text{-}OH^{-1} + Wastewater\ (Cl^{-1})\ldots \qquad (6.8)$$

6.2.9 Fuel gas

Fuel gas, which is composed of $C_{1\text{-}4}$ molecules, is produced from different streams of the plant. These waste streams have hydrogen sulfide (H_2S) gas; so the fuel gas cannot be used as is in the furnace and fire heater operations. For these reasons, a fuel gas sweetening process is used. In the process, the waste gas from the plant is sent through a knockout suction drum, where any liquid carryover is removed, then compressed and cooled, and further sent through a knockout discharge drum to remove any further liquids, before being sent to the absorber or the contactor tower. The contactor tower selectively removes H_2S present in the waste gas using the commonly used amine, such as methyl diethyl amine (MDEA) or diglycol amine (DGA). The amine rich in H_2S, also called rich amine, is sent to the regenerator, where it is boiled, and the acid gas is produced. The acid gas produced further goes to the sulfur recovery unit (SRU). It is important to note that different parameters are observed in the fuel gas treatment process at different locations. For example, the flow and composition are measured in the contactor's inlet stream to ensure that a desired flow rate is sent to the unit as per the specification. Also, the flow and the H_2S content in the sweet gas are monitored from the contractor's overhead line to determine the adequacy of the amine rate.

6.2.10 Hot oil

Hot oil is used in the plant where high-temperature users need high temperature for reboiling or when the steam supply is not available at the point. The cost study is typically done before deciding on steam versus hot oil options for grassroots plants. Unlike the steam system, the hot oil system is a closed-loop and self-sufficient system. Moreover, depending on the specific temperature requirement of the process, varieties of hot oil fluids can be chosen to meet the requirement. Fluid properties, such as viscosity, thermal conductivity, and density, are often compared to determine the correct choice of application. The hot oil circuit involves a surge drum, a fired heater, a set of circulation pumps, pump inlet basket strainers, and other process exchangers, which needs hot oil. The surge drum provides the required expansion volume for the system. As the hot oil is reused in the closed loop, it could break down over time, and a solid could form over the period. For this reason, basket strainers are used to remove any solids.

6.2.11 Tempered water

Tempered water is very similar to the cooling tower, but the temperatures in the tempered water are very specific to the process plant operation. Like the hot oil and the cooling water systems, the tempered water system involves a set of pumps, water or air type coolers, a buffer tank, and a piping network.

6.2.12 Chilled water

Similar to the hot oil or tempered water system, a chilled water system is used for very cold-temperature applications in the process. A chilled water system is a closed-loop system that involves a chiller, a chiller water storage tank, a set of pumps, cooling water exchangers, a refrigerant, and a chilled water user.

6.3 Isolation of utilities from process streams or equipment

Utilities such as steam or nitrogen are often connected to the process pipelines or equipment for purge-out or clean-out during shutdown and commissioning; low-pressure steam, which is at 50 PSIG, and low-pressure nitrogen, which is available at 30 to 40 PSIG, are used for such applications. It is necessary to separate out the process pipeline or the equipment from these utilities to avoid any backflow of hazardous chemicals or flow of hydrocarbons back into the utility system. If the hazardous chemicals or hydrocarbons are backed into the utility systems, these can create potential safety hazards for other process unit applications. The piping class of utility is always different from the process pipeline or the equipment; so it is necessary to provide the pipe specification break, right before the isolation valve. A set of check valves is always used to avoid any backflow, followed by a spectacle blind in a closed position. Between the spectacle blind and the equipment, another isolation valve is added to ensure positive isolation, and a drain connection is added to the check valve. To avoid buildup of excess hazardous chemicals inside the piping, it is always recommended to minimize the piping distance from the first isolation block to the equipment, as shown in Figure 6.2. Different customers may have other variations in the setup of utility connection to the process section. A correct design is required to avoid any safety hazards or unsafe scenarios.

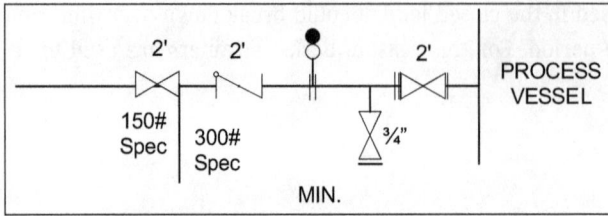

Figure 6.2: Example of a utility connection to the process equipment.

6.4 Utility stations

A typical utility station has four utilities. These utilities are clarified water, plant air, low-pressure steam, and low-pressure nitrogen. Clarified water is used for general cleaning purposes. Plant air is used to supply breathing air when a vessel entry is required, or it could be used to operate torquing tools. Low-pressure steam is used to steam-out and for cleaning purposes. Low-pressure nitrogen is used to purge a vessel or an equipment of hazardous chemicals or hydrocarbons. Depending on the need for utilities, one can have a combination of utility systems, for example, water-air, air-nitrogen, nitrogen-steam, or water-air-steam. Sometimes, hot water may be necessary for special cleaning purposes. The hot water can be produced by mixing low-pressure steam and clarified water or utility water. A detail of the utility station is explained in Section 11.21.

6.5 Utility distribution

Different utilities are usually routed in the main header of the process unit. They are branched out from the main header to the sub-headers and the different units. Each unit at the battery limit has a block valve, a pressure gauge, a transmitter, and a flow transmitter. Pressure and flow transmitters are required to report the usage from the plant. Field pressure gauges are used to check if the required amount of pressure is available in the headers. Each main header has a block valve to isolate either a single user or a group of users from the main header. A very commonly used distribution method is explained using a cooling water network distribution example in Section 11.33.

6.6 Future utility demand

Utility connections are always critical for current and future projects. It is always recommended to size the main header to account for future expansion needs. An expansion possibility of 20% to 30% of the design margin on the required capability is adequate. If a particular utility is oversized, one of the pumps or boilers can be turned off or turned on to meet changes in the process demand.

Sometimes, clients or customers provide guidance on adding valves or flanges for future expansion considerations. For example, Unit A is currently engineered by an engineering contractor, and there is a plan to build unit B five years from now. In such cases, additional valves and flanges are very beneficial. Utility lines that account for future expansion considerations are always needed. For example, for a grassroots plant, originally, the requirement for high-pressure steam was around 210,000 pound per hour, and the decision from the customer was made to install a boiler that can produce high-pressure steam at a rate of 400,000 pound per hour. So, this extra boiler capacity is not wasted but could be utilized for future expansion projects, there would be no need to add another boiler.

Chapter 7
Isolation and drain/vent valve details development

After the development of equipment, process instrumentation, and utilities, the process engineer can start adding isolation, drain, and vent valves as needed. It is easier to add valves system by system and an engineer can yellow highlight the system as each of the systems is completed. Once the process side of the design is completed, the process engineer can add valves and drains to the utility system. It is also recommended that a common drain and vent valves are added as required for the local system, such as the control valve. The control valve needs drain valves upstream and downstream.

7.1 Introduction and need for valves

There are varieties of valves available in the market, which are manufactured by several manufacturers. Valves differ in application: temperature, pressure rating, size, etc. All these valves discussed in this section are manual-type valves with no actuators or instrumentation attached to them. The valve body itself is made of metallic components. The gate acts as a restriction device and determines if the wall needs to be closed or opened. Packing is added to eliminate hazardous material leaking from the process system into the atmosphere. The gate is attached to the stem, which has grooves. The hand wheel is placed on the stem, which can be rotated to adjust the valve opening. Larger valves have larger wheels and are difficult to operate. Special tools are developed in the process industry to operate such big valves.

There are several manufacturers of valves that have a proven successful history. Even though the manufacturer is not mentioned on the P&ID, a process engineer working on a P&ID should understand the need for different valves in various applications.

The size of the valve varies from half an inch to a large bore of the size of 52 inches. The size of the valve depends on the size of the upstream or downstream piping. The pipe size depends on the flow rate through the pipe and the fluid type. Larger the flow rate, the bigger the required pipe size, and vice versa.

7.2 Valve types

7.2.1 Gate type

Gate-type valves are very commonly used in the industry because of the low cost and lower pressure drop offered by the valve. These valves are used for gas and liquid services throughout the industry. The valves are commonly used for isolation and on-

https://doi.org/10.1515/9781501519864-007

off services. Such services are widely available in the plant. Valves come in all sizes as required for the plant. Due to the type of gate design, they provide very less restriction to the flow, further providing a very low-pressure drop across the valve. They can also be used for applications where tight sealing is required. They can also be used for high-temperature and high-pressure applications and are available in all piping classes. These valves cannot be opened and closed very easily, and it may take some time for a bigger valve size for example, 30 inches. They often go through maintenance problems such as valve leaking through the packing, or the internal mechanism of the gate is not great. Also, these valves are prone to high vibrations. During the 3D modeling, it is important to ensure that any operator can access the handwheel without any safety issues. Manufacturers produce the gate valve in different metallurgies as required by the customers.

7.2.2 Globe type

Globe-type valves are also used globally in all the process plants. The difference between the globe and the gate valve is that the globe has a disc shape mechanism that stops the flow from going forward. Also, the flow direction changes while the fluid goes through the body opening. This allows the globe valve to take an excess pressure drop compared to the gate valve. The globe valve is often used to control the flow or as a substitute for the control valve when the control valve is not functioning. They provide tight shutoff as some processes may require. The major advantage of the globe valve as compared to other valves is that it does not leak. Globe valves can be easily opened or closed compared to the gate valves, which takes a lot of time. Manufacturers produce globe valves in different metallurgies as required by the customers.

7.2.3 Butterfly type

A butterfly valve is designed based on a metallic disc perpendicular to the valve's body. They do not provide a tight shutoff and good controllability, but they are inexpensive and can be used for on-off applications where flow control is not critical. They are widely used in the pharmaceutical, oil and gas, and refinery industries. Butterfly valves produce at least a small amount of pressure drop when the disc is fully open and when the flow path is completely open and only restricted by a small portion of the disc. They come in all sizes from 2 inches to 30 inches. The big butterfly valves are operated using a gear mechanism.

7.2.4 Ball type

Ball valves are widely used in oil and gas applications where reliability and pressure drop are our primary concerns. The ball valve provides good reliability and low-pressure drop in the system. They are not throttling valves, so are similar to gate valves. They can be used as an on-off operation. The valves are widely popular for applications ranging from temperature of 210 °C and pressure of 750 bar. Similar to the butterfly valves, a ball with a hole is placed inside the valve assembly. When the valve is fully opened, the open area of the pipe is very similar to the open area of the ball of the valve. They come in all sizes, from 2 inches to 52 inches, and the big valves are operated using a gear mechanism similar to the butterfly valve. Manufacturers produce the ball valve in different metallurgies as required by the customers.

7.2.5 Needle type

Needle valves have a small needle-like metal, providing a very tight shutoff. These valves are very popular in the industry as they are very reliable. One of the down-sides of the needle valves is that they can only be used for very small flow rates, such as sample stations and instrument tubing. Needle valves are made of stainless steel and can also be used for high-temperature and high-pressure applications.

Table 7.1 shows different types of commonly used valves in the industry with their typical symbols. The symbol could change from customer to customer, depending on industry standards.

Table 7.1: Types of manual valves.

Valve type	Sizes available, inches	Pressure drop	Symbol	Application purpose
Gate	2–36 commonly produced. 42–66 are specially made	Low		Isolation
Globe	2–10 are common. 12–18 are used very rarely	High		Regulation of flow
Butterfly	2–60 are generally used. Valve up to 96 inches are available depending on application.	Very low		Isolation
Ball	¼ to 60. Larger valve sizes are gear type.	Very low		Isolation and regulation
Needle	1/8 to 1	Significantly high		Regulation of flow
Pinch	1–28	High		Isolation and regulation

Table 7.1 (continued)

Valve type	Sizes available, inches	Pressure drop	Symbol	Application purpose
Plug	3–24 are common. Valves up to 72 inches are based on applications.	Low		Isolation and regulation
Diaphragm	½ to 4 are common. 6–12 do not withstand full vacuum.	Very high		Regulation of flow in corrosive media

7.3 Double block and bleed arrangement

Double block and bleed refers to an arrangement of two block valves located in very close proximity and a bleed valve in between to remove any trapped pressure when the block valves are closed. The two block valves are similar in size, and the bleed valve is very small. The purpose of the bleed valve is to vent or drain the liquid or gas trapped between the isolation valves. Double block and bleed arrangement is also called DBB and is commonly used in very high pressure and hazardous chemical applications to isolate one piece of equipment or piping from the other. Below are some examples that show the need for a DBB arrangement.

Example 1: Hazardous chemicals such as H_2SO_4

In the example shown in Figure 7.1, suppose the purpose is to hand over a pump for preventive maintenance to the mechanical department. The system should be completely drained and isolated from the hazardous chemical H_2SO_4. To do that, first, the two block valves shown are closed in the pump suction (6" valves) and discharge (4" valves). The closing action of the valves ensures that the pump is completely isolated from the H_2SO_4 tank and downstream system. Then, a bleed valve (3/4") in between the block valves can be opened, and the trapped H_2SO_4 can be drained to a safe location. Closing the valves and bleeding the trapped liquid ensures that the H_2SO_4 from the tank is positively isolated from the pump. The operator can drain the liquid from the casing to the closed drain system, add spectacle blinds (not shown in the Figure), and safely hand over the pump to the mechanical department for any possible maintenance activities.

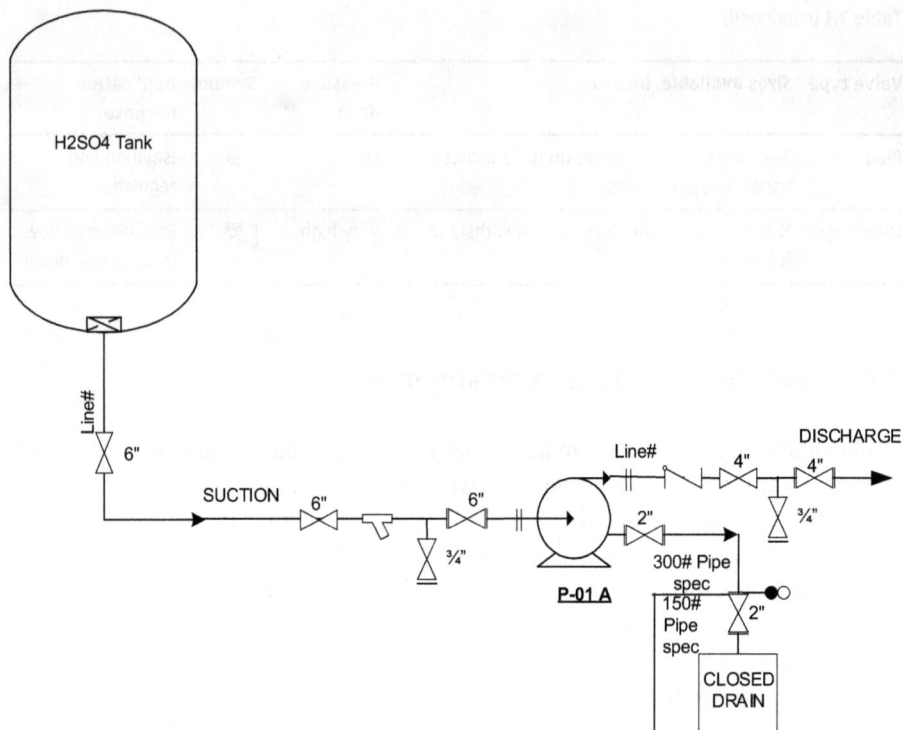

Figure 7.1: Example of double block and bleed arrangement for hazardous chemicals.

Example 2: High-pressure application for positive isolation

In this example shown in Figure 7.2, let us consider that we need to hand over the compressor to the maintenance for preventive maintenance. It is suspected that the compressor is not functioning properly. In such cases, it is important to ensure a positive isolation between the knockout inlet drum and the compressor. To do that, first, the block valves at the inlet and outlet of the drum are closed (four 18" valves). The relief valve bypass is opened (8" valve), and the vessel is completely depressurized. Once depressurized, the bleed valves (3/4") between the 18" block valves are opened, and any trapped pressure between the block valves is vented out. Also, the trapped pressure inside the compressor up until the second isolation block valve (18" located in the drum outlet) is vented by the DBB valves available (not shown in Figure 7.2) at the compressor. Spectacle blinds are flipped to the closed position. Since the compressor is positively isolated from the high-pressure knockout drum and completely depressured, maintenance can now work on the compressor as required.

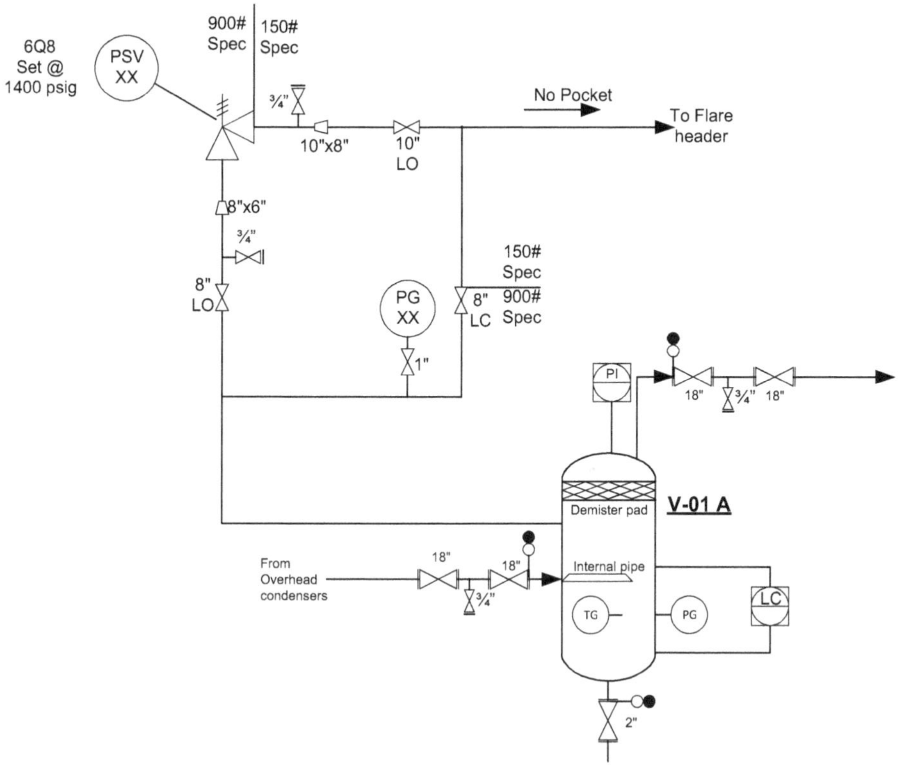

Figure 7.2: Example of double block and bleed arrangement for flammable gases.

7.4 Drain and vent valves

Most of the piping and equipment in the industry are often subjected to routine and preventive maintenance. It may be necessary for the maintenance team to overhaul the equipment. It is also a common practice to perform a through preventive maintenance of the instruments, and all the instruments routinely go through failure mechanism checks and also need routine maintenance by the instrumentation engineers. For all these purposes, isolating a piece of piping, equipment, and or instrument is necessary. For all these reasons, draining or venting the fluid inside the piping, equipment, or instrument is necessary. Draining and venting can be achieved by drain valves available on a given system. Drain and vent valves are very common and available throughout the process unit for all the processing plants, irrespective of the manufacturing process. The drain and vent valves come in sizes 1 inch, 1–1/2 inches, and 2 inches. The valves also come in different types of classes and different metallurgy depending on the requirement.

The drain valve refers to draining a section of a pipe, vessel, or equipment with liquid contents inside. The liquid contents can be hazardous or non-hazardous. If it is non-hazardous, like water, the drain can be provided with a blind flange. The non-hazardous liquid can be drained near the valve without any safety issues. On the other hand, if the liquid is hazardous, for example, benzene, an open drain cannot be achieved due to safety concerns. For such hazardous chemicals, a closed drain system is designed. This is discussed with an example in Section 11.14.

Similarly, vent valves are used to remove these gases when gas or vapor is present in the piece of piping or a vessel. If the gas is non-hazardous, for example, air, it can be vented to the atmosphere without any safety concerns. For all non-hazardous gas or vapor applications, the vent valve is provided with a blind flange. When vented into the atmosphere, hazardous gases or vapors, such as propane gas, can create a hazardous environment. For such hazardous gaseous applications, a closed-loop vent system is always designed, and venting of the hazardous gases is done using these vent valves. Positive isolation of the vent relief system from the equipment or piping is necessary to avoid creating any unsafe or hazardous situations.

High-point vents (HPV) are used to remove any trapped air during the hydrotest operation. A low point drain (LPD) removes any liquid present at the low point in the system. The system's design creates a low point, which can accumulate hazardous or non-hazardous chemicals. This can create corrosion, which leads to unsafe scenarios. These HPV and LPD valves are identified during phases three and four of the design of the plant and reviewed by the process operator of the unit during the 3D model reviews.

7.5 Special type of valves

7.5.1 Check valve

A check valve is used for applications where the backflow of the high-pressure system into the low-pressure system needs to be avoided. For example, pump discharge is always provided with a check valve to ensure that, if the pump were to fail, the outlet contents of the pump are not returned to the pump suction. Two check valves might be needed in series for severe and critical backflow prevention scenarios, and both check valves are assigned a safety criticality rating. Highly critical check valves are often taken out of service for inspection and checks to ensure the function is adequate.

The internal parts of the check valve are very simple. It involves a disc in the middle, which moves away when there is a flow through the pipe. As the flow through the pipe stops, the disc falls back to its original position based on gravity. For such reasons, the check valves are always located horizontally to ensure the disc is always operational. Swing type of check valves are common in the industry and widely used.

7.5.2 Choke valves

Choke valves are primarily installed on the gas or oil wellheads. When the oil or gas comes from the high-pressure reservoir, the Christmas tree diverts the high-pressure fluids to a choke valve. The choke valve controls the flow rate and reduces the pressure from the reservoir to the operating pressure of the oil and gas processing plant. It is necessary to control the high pressure from the oil or gas reservoir to prevent any damage to the equipment, which are rated for a much lower pressure than the oil or gas reservoir. Choke valves are highly subjected to erosion due to high velocities caused by the high-pressure differential across the choke valve. Hence, the choke valves are subjected to high maintenance in the oil and gas industry.

7.6 Blinds

Even though isolation valves are available in the system to provide a positive isolation from the high-pressure to the low-pressure or isolating the hazardous chemical from the equipment, which has gone for maintenance, the valves alone are not sufficient to provide positive isolation from the high-pressure or hazardous chemical exposure. For all such reasons, blinds are always added and used during normal operations and shutdowns. Three types of blinds are available in the industry. They are shown in Table 7.2.

Table 7.2: Types of blinds.

Blind type	Function	Sketch of a blind	Symbol on P&ID
Spacer	To fill the space between the flanges and the second flange to seal the flanges against hazardous leaks. The spacer ring is also used to create the space required when the parallel blinds need to be installed.		
Spectacle	The spacer portion of the blind helps fill the gap between the two flanges to protect against hazardous leaks. And the closed position of the blind provides positive isolation.		
Paddle	They provide positive isolation from one system to the other system.		

Chapter 8
Additional information for effective design of P&IDs

Once the engineer has added the equipment, equipment with piping, additional instrument details, the necessary isolation, and drain/vent valves, it is necessary to add details on the PI&Ds such as design notes and project notes. It is important to show additional details such as sample station and pump vents on P&IDs. All this information makes the P&IDs more complete.

8.1 P&ID legends

The legend or a nomenclature P&ID is very common and very specific for an operating company or a processing plant. The legend sheets show the details, such as insulation symbols for the piping, line numbering, service codes, example of equipment, instrument numbering, etc. These details are very important to the users to either modify or create a new P&ID as required for the project. The user must ensure that they adhere to this nomenclature consistent with P&ID standards. The user can either modify or add a new detail on the new project if required. For new projects, an engineering company may have to create a new legend sheet based on an agreement with an operating company. Details of nomenclature on a P&ID are explained in detail in the following sections.

Customers may provide the equipment, instrument, and P&ID numbers by reserving the numbers through the customer's online portal. If the numbers are not provided, an engineer developing P&IDs could use the guidelines discussed in sections below.

8.1.1 Unit numbering

The customer determines unit numbers for a new unit at the plant site based on adjacent units and the availability of a unique number. Figure 8.1 shows an example of unit numbering on a plot plan drawing. As seen, the new unit number for the proposed unit can be 03 or 02, as the numbers are available.

8.1.2 P&ID numbering

Process P&IDs can be numbered 1 to 75 if there are 250–300 equipment, assuming 3–4 equipment per P&ID sheet. Prefixes such as T-XX for a tank, V-XX for a pressure

https://doi.org/10.1515/9781501519864-008

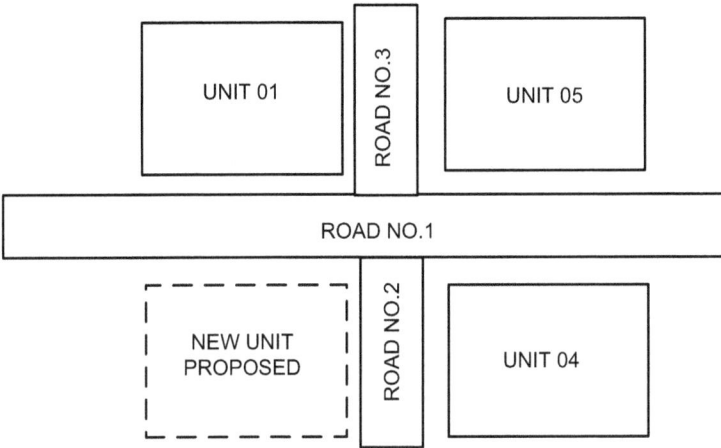

Figure 8.1: Unit numbering on a plot plan drawing.

vessel, R-XX for a reactor, E-XX for an exchanger, P-XX for a pump, etc., are commonly used in the industry. Based on this guide, assuming the first major equipment on the first page of a P&ID is a storage tank. The equipment number for the tank becomes T-100. If there are pumps on these drawings, their numbers would be P-100, P-101, etc. Any second tank on this drawing would have an equipment number of T-101. Also, considering a plant name to be AKUS, the area number of the unit can be A1, and the process area is primarily denoted by P (U can denote utility area). The P&ID number is denoted as: plant name-area number of a unit-process or utility area-major equipment number without prefixes. The P&ID number becomes AKUS-A1-P-100.

Similarly, let us consider that the major equipment for a second P&ID sheet is an exchanger, and it is a process P&ID. The tag number for the exchanger becomes E-200. The P&ID number becomes AKUS-A1-P-200. If there is no equipment on the P&ID and a P&ID is needed to show several pipes (e.g., battery limit drawing), the next equipment number can be chosen after the completion of numbering all equipment in the unit. For example, let us say that the equipment number of the 800 series is the last equipment counted in the unit; the battery limit drawing number becomes AKUS-A1-P-900.

8.1.3 Pipeline codes

Pipeline codes are commonly specified as: D-4"-ABCD-100-3055-PMC-1"-STR, where:
- D: diesel (process service code)
- 4": line size
- ABCD: pipe specification
- 3055: line number

- 100: Unit number
- PMC: Material of construction of pipe (could be process or utility)
- 1": thickness of insulation
- STR: steam type of tracing

8.1.4 General piping insulation types

Some or most of the piping in each plant has some kind of insulation to protect the contents in the pipe from heat loss. This insulation can be used for heat conservation, personal protection, heat tracing, etc. These insulations are shown along with the pipeline codes. Below are some commonly used insulation codes in the industry and their codes:

Heat conservation: HEC

Personal protection: PEP

Steam tracing: STR

Electric tracing: ETR

Fire protection: FIP

8.1.5 Process service codes

Any process fluid that is not a utility is called a process. The service codes for each fluid differentiate them from one another. Process service codes are part of the pipeline codes. Below are some examples of the process service codes. Note that the project or the plant can customize the codes per their products. Care should be taken to avoid duplication of the code. For example, propane and pentane can be written as "P" but could conflict with one another. Codes such as "P" and "PE" can be used to be unique. Table 8.1 shows different service codes commonly used in the refinery.

Table 8.1: Different service codes used in the refinery.

Raw crude: RC	Propane: P
Hydrogen: H	Propylene: PR
Urea: U	Ethane: E
Diesel: D	Ethylene: ET
Kerosene: K	Butane: B
Vacuum gas oil: VGO	Pentane: PE
Heavy gas oil: HGO	Hexane: HE

8.1.6 Utility service codes

Utilities are used for all industries and all types of plants. Utility service codes are part of the pipeline codes, similar to process service codes. Some common examples of utility service codes are shown in Table 8.2.

Table 8.2: Utility service codes.

Instrument air: INA	Hydrogen fuel: HFU
Plant air: PLA	Oxygen: OX
High-pressure steam: HPRS	Natural gas: NG
Medium pressure steam: MPRS	Fuel oil: FO
Low-pressure steam: LPRS	Fuel gas: FGA
High-pressure condensate: HPRC	High-pressure nitrogen: HPRN
Medium-pressure condensate: MPRC	Low-pressure nitrogen: LPRN
Low-pressure condensate: LPRC	Demineralized water: DWA
Boiler feed water: BFEW	Filtered water: FWA
Cooling water supply: CWAS	Softened water: SWA
Cooling water return: CWAR	Chemicals: C
Utility water: UWA	Chilled water supply: CHWS
Potable water: PWA	Chilled water return: CHWR

8.1.7 Instrument numbering

Several customers have different methods of producing instrument numbering. The following guidelines are provided as an example.

Suppose the equipment number on a P&ID drawing is T-100 (T represents a tank), and the area of the unit number is 2. The instrument numbering would be Instrument type-2-100-01 (01 represents the 1st instrument). The following instrument tags are shown as an example:

- PT-2-100-01 (Pressure transmitter number 01 located in unit 2 on equipment T-100)
- PT-2-100-02 (Pressure transmitter number 02 located in unit 2 on equipment T-100)
- TG-2-100-01 (Temperature gauge number 01 located in unit 2 on equipment T-100)
- TG-2-100-02 (Temperature gauge number 02 located in unit 2 on equipment T-100)
- LT-2-100-01 (Level transmitter number 01 located in unit 2 on equipment T-100)
- LT-2-100-02 (Level transmitter number 02 located in unit 2 on equipment T-100)

8.1.8 Specialty items

Specialty items (denoted by SI) cannot be categorized into equipment or instrument and are categorized as specialty-type items. Some commonly used specialty items are blinds, atmospheric vent silencers, expansion joints, hoses, safety showers, steam traps, desuperheaters, special valve types, etc. If the specialty item size (commonly the same as upstream and downstream pipe size) is 1", then the item tag becomes SI-01-1." The following sequence numbering can be followed: SI-01-size.

8.1.9 Valve type and symbols

Other types of valve symbols are shown in Table 8.3.

Table 8.3: Common valve types.

Type of valve	Symbol on P&ID
3-way	
Choke	
Orbit	
Pinch	

8.1.10 Equipment types

Different equipment are shown differently on the P&IDs to differentiate their function from one another. Different types of equipment are shown in Table 8.4.

Table 8.4: Equipment types and symbols.

Equipment	Equipment symbols on P&IDs	Equipment	Equipment symbols on P&IDs
Horizontal centrifugal pump	 OR 	Storage tank – fixed roof	

Table 8.4 (continued)

Equipment	Equipment symbols on P&IDs	Equipment	Equipment symbols on P&IDs
Horizontal centrifugal pump right		Storage tank – floating roof tank	
Positive displacement pump		Storage tank – sphere	
Vertical in-line pump		Column with packing	
Metering pump		Vertical pressure vessel	
Progressive cavity pump		Reactor	
Screw pump		Horizontal pressure vessel	
Vertical can pump		Distillation tower	

Table 8.4 (continued)

Equipment	Equipment symbols on P&IDs	Equipment	Equipment symbols on P&IDs
TEMA AEL-type exchanger		Centrifugal compressor unit	
TEMA AEM-type exchanger		Reciprocating compressor unit	
TEMA AEU-type exchanger		Screw compressor unit	
Double pipe exchanger-type exchanger		Electric motor driver	
TEMA BKU-type exchanger		Turbine driver	
Air -cooled exchanger-type exchanger	E-01 A	Mixer or agitator	

8.1.11 Pipe fittings and other pipe symbols

The blind types are mentioned in Table 7.2. Other types of pipe fittings that include blind flanges, reducers, strainers, filters, silencers, eductors, ejectors, etc. The types of fittings and their symbols are shown in Table 8.5.

Table 8.5: Pipe fittings and other pipe symbols.

Type of fitting	Symbol
Blind flange	
Cap	
Concentric reducer	

Table 8.5 (continued)

Type of fitting	Symbol
Flat bottom reducer	
Sample connection	SC
Strainer – Y-type	
Strainer- temporary	
Strainer – basket-type	
Filter	
Spool –removable	ROS
Vent silencer	V S
Ejector	
Eductor	
Corrosion coupon	CC
Requirement of a slope	1:100
Free drain using gravity towards the direction shown	FREE DRAINING

Table 8.5 (continued)

Type of fitting	Symbol
Open drain system	
Closed drain system	CD
Piping specification break	PIPE SPEC A PIPE SPEC B
Piping flexible hose connection	
Hose connection	
Swing elbow	
Steam trap	STR
Desuperheater	DES

8.1.12 Scope legend

It is important to differentiate between two types of scope, e.g., new scope from existing piping, so the project and planning can be executed appropriately. Types of scope clouds are mentioned in Table 8.6.

Different scope clouds around a piece of piping and their meaning are shown in Table 8.7.

Instrumentation symbols can vary from customer to customer and project to project. The most common types of instruments are shown in Table 8.8.

Table 8.6: Different scope clouds, in general, used in the P&IDs.

Cloud type	Symbol	Meaning
New scope post-turnaround		New scope post-turnaround for equipment, piping, and instrumentation is shown in the solid cloud.
New scope pre-turnaround		A scope pre-turnaround for equipment, piping, and instrumentation is shown in the double-dotted cloud.
Existing as-built		Any corrections to the existing system can be shown in a single dotted cloud.
Vendor scope boundary	VENDOR SCOPE	Vendor scope is shown in the dashed box with anticipated in and out piping connections.

Table 8.7: Different scope clouds around piping used in the P&IDs.

Line scope	Symbol	Meaning
New piping above ground		New above-ground piping shown in solid cloud
New piping underground		New below-ground piping shown in solid cloud
Differentiation of new piping from existing	TP XXXX	Tie-in of new above-ground piping attached to existing pipe depicted by a solid cloud
Future pipe		Any possible future piping shown in a dotted cloud

Table 8.8: Other instrumentation symbols.

Instrument item	Symbol	Meaning
Controller		Provides control function
Interlock		Provides emergency shutdown based on a predetermined set value
Field		Instruments located in the field

8.2 Design notes

Design notes are required all the time. A design note conveys an essential message to the P&ID user about the equipment, piping, or instrument, which may not be evident from the P&IDs. Not all notes are to be present on the P&ID. P&ID is an engineering document, and careful consideration should be given while adding a design note to the P&ID. The design notes can be related to the equipment design or safety of a plant or equipment. Table 8.9 shows different design notes applicable to different parts of a process unit.

Table 8.9: Examples of design notes.

Note classification	Actual design note	P&ID representation of the note	Reason for the note
Design	Note 1. High point vent and low point drains.		Designations used to define the purpose as explained in Section 7.4
Design	Note 2. Locate the valve LCV-001 close to the drum D-001.		To minimize vibration effect due to two-phase flow and flashing

Table 8.9 (continued)

Note classification	Actual design note	P&ID representation of the note	Reason for the note
Safety	Note 3. Install the check valve in a horizontal position.	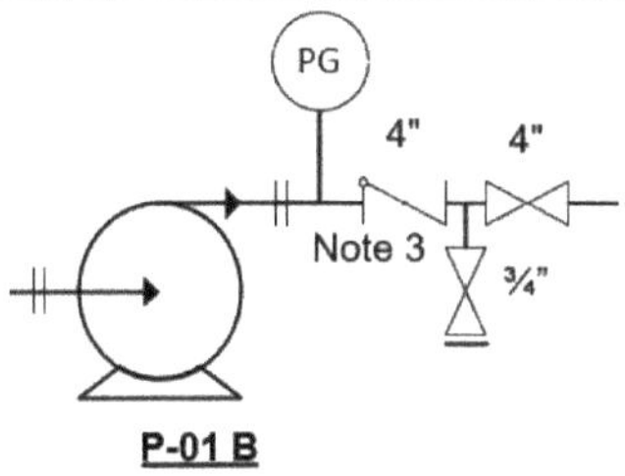 PG 4" 4" Note 3 ¾" **P-01 B**	To avoid malfunction of operation when vertically installed
Safety	Note 4. Route the discharge piping to a safe location.	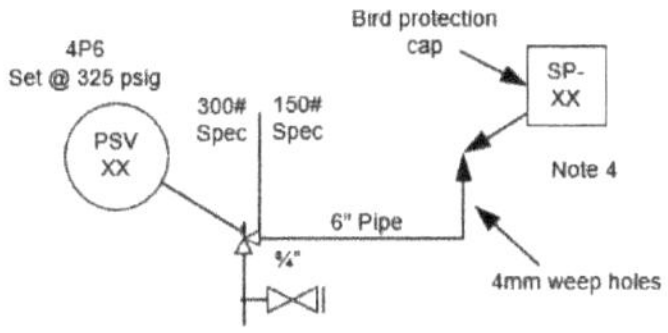 4P6 Set @ 325 psig PSV XX 300# Spec 150# Spec Bird protection cap SP-XX Note 4 6" Pipe ¾" 4mm weep holes	To avoid hot steam hurting nearby personnel
Design	Note 5. Provide five pipe diameters upstream of FT-001.	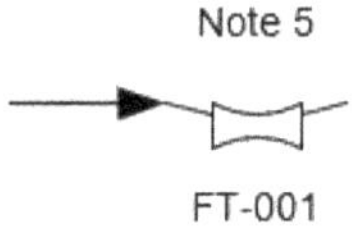 Note 5 FT-001	For proper functioning of the instrument measurement
Design	Note 6. Provide top connection for all steam consumers.	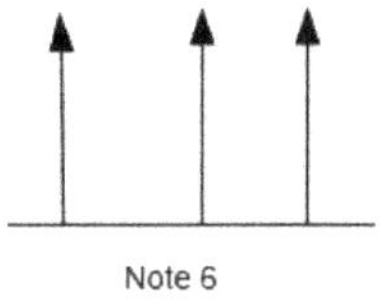 Note 6	To avoid condensate entry to the users
Safety	Note 7. Locate steam trap discharge at a safe location.	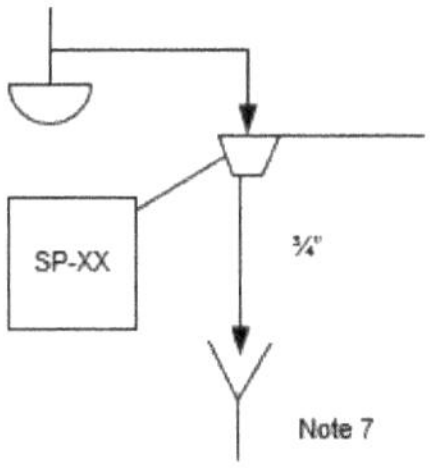 SP-XX ¾" Note 7	To avoid plant personnel getting hurt from hot water and steam

8.3 Project notes

Project notes are added on the P&IDs to alert other disciples about the unfinished work or a follow-up about some piping or equipment. Table 8.10 shows examples of such project notes on P&IDs.

Table 8.10: Examples of project notes.

Project note number	Project note detail	P&ID representation of the note	Reason for the note
Project Note 1 (PN1)	Heater vendor is to be finalized.		Note added to alert other piping and instrument engineers that the project engineer has not finalized the vendor selection of the heater
Project Note 2 (PN2)	Closest natural gas tie-in from the boiler is to be located.		Note added to alert piping engineer that a closer tie-in is possibly needed to save the cost of the project

8.4 Hold notes

Hold notes on a P&ID show that the design of a piping, instrument, or equipment is missing some information, and hence the design cannot be completed. These notes are used to notify other disciplines about the missing information. Table 8.11 shows several examples of hold notes and the reasons for their use.

Table 8.11: Example of hold notes on a P&ID.

Hold note number	Hold note detail	P&ID representation of the note	Reason for the note
Hold 1	Hold for safety valve design		Note added to alert other piping and instrument engineers that the process engineer has not finalized the size of the safety valve
Hold 2	Hold for control valve size		Note added to alert cost estimating that the instrumentation engineer has not finalized the size of the control valve
Hold 3	Hold for piping size		Note added to alert the piping engineer that the process engineer has not finalized the pipe size.
Hold 4	Hold for vendor information		Note added to alert everyone on the project team that the mechanical engineer is waiting for further vendor information on a piece of equipment
Hold 5	Hold for the customer's approval		Note added to alert everyone on the project that a critical piece of piping needs further approval from the customer before proceeding further on the design

8.5 Sample station details

A specific sample from the processing plant is collected and sent to the laboratory for analysis. The laboratory analyzes the composition of the fluid and later posts the analysis on the company's internal website. Sample stations are needed to check the product specifications from time to time. The control system operator looks at the sample and takes necessary action in the processing plant to meet the product specification. For example, for a benzene and toluene column separation system, the product specification of benzene can tolerate only 100 parts per million (ppm) of toluene in benzene. Suppose the sample analysis shows that the benzene product has more like 2,000 ppm toluene. In that case, the operator increases the column pressure, increases the reflux, or decreases the column overhead temperature to minimize the carryover of toluene in the overhead column system.

A typical sample station requires a pressure differential. The sample can be taken across a pump system and across a filter. Sometimes, the sample may be hot and needs to be cooled before it can be safely stored in the sample bottle. For this reason, a small cooler is added, which is cooled by cooling water. The gaseous samples are critical as, during sampling, the operator could lose some of the gas composition. Hence, the gas samples are collected into the pressurized sample vessels. These samples are collected in an enclosed sample chamber for very volatile and hazardous samples, and any volatile material that is not collected is vented out to a safe location.

Sample stations are connected by isolation valves that can be used for the isolation. A drain valve to remove any previous samples, connecting small tubing, and local pressure and temperature gauges for the operators to see the process parameters of the

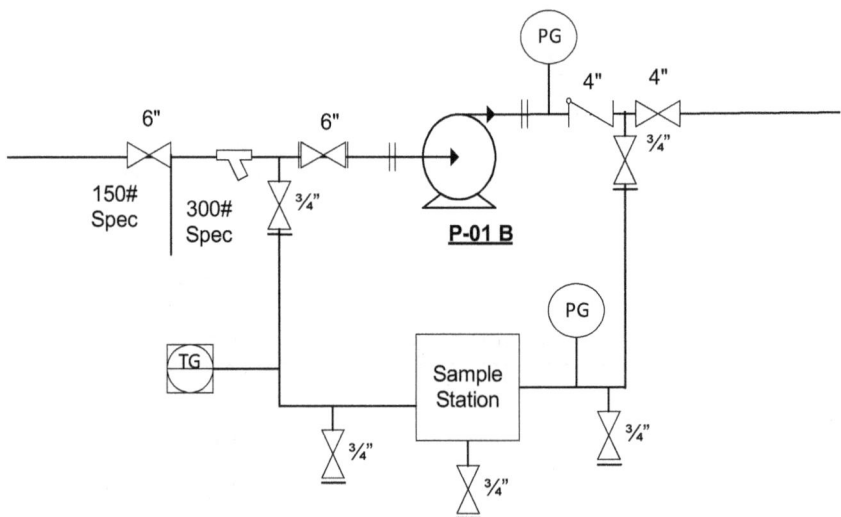

Figure 8.2: Sample station.

sample being collected visually are also available. Figure 8.2 shows an example of a sample station located across the pump.

8.6 Pump seal plan details

Every pump has flanges, gaskets, and a few mechanical components that can leak. Depending on the material handled by the pump, the material can be hazardous or non-hazardous. If it is non-hazardous such as water, the leak can be released into the atmosphere without any concerns. If the material is a hazardous chemical, leak detection is important as the hazardous material cannot be released into the atmosphere. For detecting the leak, most of the pumps in the chemical plants and refining industries have a seal plan. There are different types of seal plans depending on the type of material the pump handles.

8.7 List of P&IDs

Every process plant has a set of P&IDs. Depending on the complexity of the process operation, the P&ID count may vary. A large number of P&IDs are expected for a very complex system (e.g., coker unit). A small number of P&IDs are the norm for a simple process unit (e.g., nitrogen unit). If it is a complex unit, it is difficult to find a particular equipment in a large stack of P&IDs. P&IDs are often required for revisiting the actual plant information for HAZOP, meetings, planning, or projects. A list of P&IDs is always created and placed at the beginning of the P&ID set for process units for easier tracking and finding information. Table 8.12 shows an example of a P&ID list.

Table 8.12: Example of list of P&ID's.

P&ID number	Service	Area
AKP-55-PD-001	Reactant feed drum and feed drying system	Reactor
AKP-55-PD-002	Reactor feed pump system	Reactor
AKP-55-PD-003	Reactor 1	Reactor
AKP-55-PD-004	Product separation drums and coolers	Product separation
AKP-55-PD-005	Gas compression	Product separation
AKP-55-PD-006	Distillation column	Product separation
AKP-55-UT-001	Steam network	Utility
AKP-55-UT-002	Cooling water tower	Utility
AKP-55-UT-003	Cooling water distribution	Utility

8.8 List of equipment by type

There are different types of equipment in a processing plant. The type of equipment depends on the manufacturing process. Typical equipment types involved are pumps, distillation columns, heat exchangers, filters, storage tanks, etc. It is difficult for a complex processing plant to locate specific equipment in a large stack of P&IDs. The list of equipment, by type, is placed at the beginning of the P&ID stack to help find particular equipment easily for meetings and planning of project scope purposes. Table 8.13 shows an example of a list of equipment, by type.

Table 8.13: Example list of equipment, by type.

Equipment tag number	P&ID number	Service
PUM-001	AKP-55-PD-002	Reactor 1 feed pump
PUM-002	AKP-55-PD-002	Reactor 1 spare feed pump
RECT-001	AKP-55-PD-003	Reactor 1
COMP-001	AKP-55-PD-005	Process gas compressor
COMP-002	AKP-55-PD-005	Process gas spare compressor
DRM-001	AKP-55-PD-004	Gas and liquid separation drum

8.9 List of safety and control valves

Similar to the list of P&IDs and list of equipment by type, it is important to locate safety and control valves in the process plant easily. A list shows the location of the safety and control valves on the P&IDs. Tables 8.14 and 8.15 show example lists of safety and control valves, respectively.

Table 8.14: Example list of safety valves.

Safety valve tag number	P&ID number	Where located
PSV-001	AKP-55-PD-001	Reactor feed drum
PSV-002	AKP-55-PD-002	Reactor 1 feed pump discharge line
PSV-003	AKP-55-PD-003	Reactor 1
PSV-004	AKP-55-PD-005	Process gas compressor outlet piping
PSV-005	AKP-55-PD-004	Gas and liquid separation drum

Table 8.15: Example list of control valves.

Control valve tag number	P&ID number	Purpose
LCV-001-01	AKP-55-PD-001	Reactor feed drum level control
FCV-002-01	AKP-55-PD-002	Reactor 1 feed pump discharge line flow control
PCV-001-01	AKP-55-PD-003	Reactor 1 pressure control
PCV-001-02	AKP-55-PD-005	Process gas compressor outlet line pressure control
LCV-001-02	AKP-55-PD-004	Level control of gas and liquid separation drum

8.10 P&ID checklist

Different sections and different types of equipment on P&IDs are always checked in all the phases of the project to make sure the quality of P&IDs is maintained at all times. For such reasons, a quality audit checklist is checked by an auditor. The process lead gets audited by an auditor based on the quality checklist. If the process lead fails to comply with some of the items on the checklist, the auditor recommends options to remove the deficiencies. The customer typically provides a checklist. The EPC industry generally uses an internal checklist for all projects. Several details in the P&ID checklist are provided in sections below.

8.10.1 P&ID in general

1. Confirm if Lead Process Engineer has a master Bluebeam session marked "Master." _____
2. IFD/IFC P&IDs are reviewed with the engineers internally? _____
3. IFD/IFC P&ID drawing index issued by the drafting team? _____
4. IFD/IFC P&IDs reviewed by the Client? _____
5. IFD/IFC P&IDs reviewed by Control Systems? _____
6. Symbols used for all instrumentation, equipment, and valves? _____
7. Equipment and line numbers are added? _____
8. All holds are noted on IFD and no holds on IFC drawings? _____
9. Master markup (MMU) P&ID items are customer approved for IFC drawings? _____
10. Existing areas of the drawings field checked by piping? _____
11. Connectors from/to the drawings defined and match with the other drawings? _____
12. Utility line from the P&ID matching with utility distribution drawing? _____
13. IFD/IFC P&IDs include these items? P&ID titles and unique numbers _____

       Equipment names and numbers   _____

       Instrumentation   _____

       All needed valves and piping details   _____

14. Equipment titles show details such as design condition,   _____
    material of construction, insulation type, thickness, etc.?
15. Required equipment elevation shown?   _____
16. Motors details adequately shown?   _____
17. Equipment insulation shown on the equipment?   _____
18. All utility connections shown?   _____
19. All nozzle sizes and connection shown appropriately?   _____

8.10.2 P&IDs: piping in general

1. Process line hydraulics done and line sizes provided?   _____
2. Process and utility lines piping specs shown?   _____
3. Process and utility lines insulation shown?   _____
4. Sizing for (>2") utility line sizes completed and shown   _____
   along with insulation specs?
5. Piping specs defined?   _____
6. Piping spec breaks shown as needed?   _____
   a. Changes in design pressure (e.g., downstream of letdown   _____
      Station)
   b. Material of construction change   _____
7. Pipe specs followed per customer requirements?   _____
8. Insulation and tracing details shown?   _____
9. Connecting arrows in/out of P&ID shown properly?   _____
10. P&IDs show all line numbers for process and utility?   _____
11. In-line valves shown and identified per customer legend?   _____
12. Instrument connections shown correctly on P&IDs?   _____
13. Special notes for process piping/utility piping shown   _____
    e.g., do not pocket, free draining, long radius bends,
    min/max distances, etc.?
14. Tie-in points shown for process and utility connection   _____
    shown per customer guidelines?
15. Line sizes for vent lines shown?   _____
16. Vent lines specs sufficient for services?   _____
17. Necessary start-up and shutdown lines shown?   _____
18. Necessary equipment vents and drains provided?   _____
19. All necessary bypass line and valves identified and shown?   _____

20. Piping insulation identified and shown? _____
    a. Heat conservation (HC) _____
    b. Personnel protection if temperature > 140 °F (PP) _____
    c. Winterization protection for no flow lines _____
    d. Process fluid corrosion protection < atmospheric dew point _____
    e. For fireproofing _____
21. Tracing provided for necessary lines? _____
    a. Steam tracing (ST) _____
    b. Electric heat tracing (ET) _____
22. Isolation valves on steam inlet and outlet? _____
23. Steam silencer for steam outlet to atmosphere? _____
24. Steam traps shown properly? _____
25. Reducers identified and shown? _____
26. Spacer and spectacle blinds shown? _____
27. Temporary strainers shown? _____
28. Double blocks and bleeds required? _____
    a. To avoid product cross-contamination _____
    b. For services > 540 °C or > 50 bar _____
    c. For control valves _____
    d. Isolating spare equipment with other in operation _____
29. Tie-ins specified correctly? _____
    a. Can the line be emptied, gas freed, and flushed? _____
30. Local pressure gauges shown? _____
    a. At pump and compressor discharges _____
    b. For DeltaP across exchangers (both sides) _____
    c. In the vapor space of vessels _____
    d. Across filters and strainer for measuring DeltaP _____
    e. At turbine exhaust _____
31. Block valves properly shown to allow maintenance? _____
    a. To isolate exchangers _____
    b. To isolate flow meters _____
    c. To isolate control valves _____
    d. To isolate spare filters _____
    e. To isolate parallel equipment _____
32. Piping large enough for thermowell? _____
    a. Pipe ≥ 3"? _____

8.10.3 P&IDs: instrumentation in general

1. In-line process and utility instruments shown?   _____
2. Process and utility control valves and regulators shown?   _____
3. Process analyzer sample details shown?   _____
4. Instrument connection shown correctly?   _____
5. Safety critical interlocks provided?   _____
6. Control loop logic and instrumentations shown correctly?   _____
7. PSV instruments such as PG shown?   _____
8. Rupture disc instruments shown?   _____
9. PSV set pressures shown?   _____
10. Control valve sizes specified?   _____
11. Automatic isolation valve shown?   _____
12. Failure positions for control valve shown?   _____
13. Control valve station details shown?   _____
    a. Bypass or a handwheel provided?   _____
    b. Bypass globe valve sized?   _____
    c. Reducers shown around control valves?   _____
    d. Bleeders shown between valves and control valves?   _____
    e. A parallel control valve necessary for the range?   _____
14. Insufficient instrumentation to operate the process?   _____
15. Critical analyzer details provided?   _____
    a. Sample point provided?   _____
    b. Impingement collector details?   _____
    c. Orientation of a sample?   _____
    d. Sample filter and steam tracing provided?   _____
    e. Sample drain or vent shown?   _____
    a. DeltaP across analyzer measured?   _____
16. In-line flow meters provided and shown correctly?   _____
    a. Temperature and pressure compensation required?   _____
    b. Vapor flow measurement done?   _____
    c. Type of correct flow meter chosen and depicted?   _____
17. Level instruments shown correctly?   _____
    a. Type of level instrument shown correctly?   _____
    b. Customer design guidelines followed?   _____
18. Level switches provided in appropriate locations?   _____
19. Level glasses shown appropriately?   _____
    a. Level glasses type shown?   _____
    b. No level glasses for highly toxic or
       hazardous services?   _____
20. Interface level shown for separator vessels?   _____

21. Dip pipes and lengths shown on vessels   _____
22. Instrument hardware and control logic shown properly?   _____
    a. Transmitter/controller/control valve?   _____
    b. Alarm off transmitter output?   _____
    c. Independent alarm for critical service?   _____
    d. Interlocks properly identified   _____
23. Temperature indicator (TI) shown?   _____
    a. TIC specified correctly?   _____
    b. TI provided for liquid portion of vessels?   _____
    c. Spaced correctly for tower temperature profile?   _____
    d. Discharge of a compressor?   _____
24. Local temperature gauges shown?   _____
    a. Exchanger DeltaT for both sides?   _____
    b. Level space in vessel and tanks?   _____

8.10.4 P&IDs: pressure safety valves (PSV) in general

1. Safety valves details shown correctly?   _____
    a. PSV located at correct location to protect the equipment?   _____
    b. "Car-seal-open" (CSO) block valves shown?   _____
    c. "Car-seal-close" (CSC) block valves shown?   _____
    d. Vents and drains shown?   _____
    e. All equipment adequately protected?   _____
2. PSV/rupture disc required?
    a. A high pressure alarm between disk and PSV?   _____
    b. PI and between disk and PSV?   _____
    c. A flow check valve shown?   _____
3. Is the PSV system shown appropriately?
    a. Routed to blowdown drum if liquid scenario?   _____
    b. Routed to atmosphere if water/steam?   _____
        i.   Steam silencer provided?   _____
        ii.   Cap provided on discharge and weep hole?   _____
        iii. Elevation above 10 ft.?   _____
    c. A separate cold relief header if the relief can freeze   _____
        in temperatures < 32 °F?
4. Is PSV shown for thermal expansion on lines?   _____
    a. Long line for solar heating?   _____
    b. Pipelines for less than ambient temperatures?   _____
    c. For steam/electric traced lines?   _____
    d. Services like NH_3, cryogenic liquids, etc.?   _____

8.10.5 P&IDs: utility systems in general

1. Steam system letdowns and relief valves shown?   _____
2. Desuperheater details shown with temperature and boiler feed   _____
    water control?
3. Pressure safety valve provided on the low-pressure side to   _____
    protect max capacity requirement?
4. Cooling coils shown with utility pipes?   _____
5. Utility station details provided?   _____

8.10.6 P&IDs: electrical in general

1. Electric controls for motors shown?   _____
2. Critical power for motors shown?   _____
3. Grounding requirements for motors specified?   _____

8.10.7 P&IDs: civil in general

1. Secondary containment area shown?   _____
2. Drain routing to a sewer shown correctly?   _____
3. Curbs details shown with spill or fire requirement?   _____

8.10.8 P&IDs for pumps

8.10.8.1 Pump: design data
1. To account for the NPSH requirement, equipment elevations   _____
    shown?
2. Vibration or overheating protection provided for the pump?   _____
3. Pump lube oil system details shown, if applicable?   _____
4. Start-up strainer available and tagged?   _____
5. API Seal system shown?   _____
6. Seal cooler provided with details?   _____
7. Important design notes shown?   _____

8.10.8.2 Pump: safety and operation
1. Insulation, size, and service specified?   _____
2. Control instrumentation (Flow or pressure) shown?   _____
3. Minimum flow control required and shown?   _____

4. Process parameters and instrumentation such as   _____
    PG, PI, FI, FT, TG, TI, etc. shown?
5. Pump sparing specified as per customer requirement?   _____
6. Valving and piping meets operating and maintenance   _____
    requirements and approved by customer?
7. Future provisions shown?   _____
8. Material of construction meets the safety standards?   _____
9. Personnel protection insulation provided?   _____
10. Backflow prevention required and shown?   _____
11. Overpressure protection considered and adequate safeguards _____
    provided?
12. Casing drain/vent for safe disposal of materials?   _____
13. Pump vents and drains are closed to avoid leaks?   _____
14. Material containment required? Provided?   _____
15. Electricity grounding/bonding provided?   _____
16. Piping spec break shown where needed?   _____
17. Eyewash/safety shower shown near the pump?   _____
18. Utility station provided near pump?   _____
19. Emergency shutdown required?   _____
20. Firewater deluge or sprinkler needed?   _____
21. Adequate isolation of spare pump provided?   _____
22. Car-seal-open (CSO) or car-seal-close (CSC) shown?   _____

8.10.8.3 Pump: motor driver

1. Instrumentation and electrical details for remote or auto   _____
    stop/start shown?
2. Interlock instrumentation shown and adequately described   _____
    in the design notes?

8.10.9 P&IDs for steam driven turbines

1. Steam turbines details provided?   _____
    a. Strainer in the steam inlet shown?   _____
    b. A safety valve provided?   _____
    c. Whistle provided?   _____
    d. Steam trap shown in inlet?   _____
    e. A control valve shown on steam line?   _____
    f. Warm-up line shown on steam side?   _____
2. Design data for the turbine shown?   _____
3. Steam inlet piping, steam trap, block valves shown?   _____

4. Steam outlet piping and valving shown? _____
5. Steam relief valve details shown? _____
6. Steam supply shutdown valve shown? _____
7. Turbine speed control shown? _____
8. Turbine lube oil system details shown? _____
9. Turbine casing vents and drains shown? _____
10. Details for steam separator shown? _____
11. Process parameters and instrumentation such as _____
PG, PI, FI, FT, TG, TI, etc. shown?

8.10.10 P&IDs for compressors

8.10.10.1 Design data
1. Buffer gas and seal oil chosen correctly for the system? _____
2. Line sizes, specs, services, insulation, etc., shown? _____
3. Suction strainer provided with a PDI across it? _____
4. Piping free drain from the compressor on both sides? _____
5. Suction line steam or electrically traced? _____
6. Surge overpressure protection provided for centrifugal? _____
7. Relief valve provided on the reciprocating compressor discharge? _____
8. Shutdown interlock logic depicted on the P&IDs? _____

8.10.10.2 Knockout drum
1. High-level shutdown and general liquid levels reflect _____
operator response time? _____
2. Vortex breakers provided for pump suctions? _____
3. Demister pad shown? _____
4. Manways shown? _____
5. Liquid flow from knockout drum constant? On/off _____
level pump controller or a straight level control provided?

8.10.10.3 Auxiliaries
1. Piping to/from compressor auxiliaries _____
adequately shown on a P&ID?

8.10.10.4 Spillback control
1. Surge protection provided for centrifugal compressor? _____
2. Spillback control valve and line sized appropriately? _____
3. Spillback line comes off from the discharge line? _____

8.10.10.5 Control

1. All TIs present for temperature compensation of flow? _____
2. Pressure measured in both inlet and outlet lines? _____
3. Flow measured in inlet and outlet? _____
4. Necessary alarms and shutdowns provided? _____
5. Product specs maintained? _____
6. Change in feed rate controlled? _____
7. Change in feed composition controlled? _____
8. Product analyzers required and shown? _____
9. Instrument engineer reviewed the P&IDs? _____
10. Surge control experts reviewed the logic? _____

8.10.10.6 Miscellaneous

1. System be vented and drained? _____
2. Sample points shown? _____
3. Future connections provided? _____

8.10.11 P&IDs for vessels

8.10.11.1 Design data

1. Equipment tag to include the following information? _____
   - Equipment number, Service _____
   - Dimensions (diameter & height/length) _____
   - Design pressure and temperature _____
   - Material of construction _____
2. Equipment symbol includes the following? _____
   - Nozzle and manway sizes _____
   - All liquid levels _____
   - Elevation of skirt _____
   - Piping trim _____
   - Insulation type/thickness _____

8.10.11.2 Process and safety

1. Line sizes, specs, service, insulation, etc., shown? _____
2. Multiple inlets required and, if so, shown? _____
3. Vortex breakers for pump suctions? _____
4. Skirt height/bottom elevation shown? _____
5. Liquid level bridle connections shown for all levels? _____
6. Demister pad shown? _____

7. Manways shown? Nozzle sizes and locations shown?   _____
8. System be vented and drained? Purge connections provided?   _____
9. Toxic chemicals routed to safe location?   _____
10. System can be started up or shutdown?   _____
11. Sample points shown?   _____
12. Vessel protected from overpressure scenarios?   _____
13. Insulation indicated?   _____

8.10.12 P&IDs for columns or towers

8.10.12.1 Design data
1. Equipment tag includes the following information?   _____
   - Equipment number, Service   _____
   - Diameter and height   _____
   - Design pressures and temperatures   _____
   - Material of construction   _____

8.10.12.2 Process and safety
1. Line sizes, specs, service, insulation, etc., shown?   _____
2. Feed location shown, distributor provided?   _____
3. Multiple feed points shown?   _____
4. Vortex breakers for pump shown?   _____
5. Skirt height shown to provide   _____
   a. Appropriate NPSH for pumps   _____
   b. Reboiler circulation?   _____
6. Liquid levels shown to allow adequate liquid residence time?   _____
7. Trays with passes and downcomers shown?   _____
8. Liquid or vapor distributors for packed bed shown?   _____
9. Demister shown, if required?   _____
10. Manways, nozzles, tags, and locations shown?   _____
11. Sufficient vents and drains shown?   _____
12. Vents and drains provided for the system?   _____
13. Purge connection provided?   _____
14. Toxic drains and vents routed to safe location?   _____
15. System can be started up?   _____
16. Sample points shown?   _____
17. Sight glasses required and shown?   _____
18. Column protected from overpressure scenarios
    and PSV shown with set point and size?   _____
19. Insulation indicated on the P&ID?   _____

20. Electrical ground required and noted?   _____
21. Minimum Design Metal Temperature (MDMT)   _____
22. mentioned? (Steamout or vacuum conditions)

8.10.12.3 Control

1. Control valves and instrument loops shown?   _____
2. Column pressure control shown?   _____
3. Column temperature control shown?   _____
4. Reflux flow control shown?   _____
5. Column bottoms level control shown?   _____
6. Product specs controller shown?   _____
7. Feed rate controlled?   _____
8. Feed composition change controlled?   _____
9. A PDI is required across the column?   _____
10. Sufficient temperature profile across the column   _____
    can be obtained?   _____
11. All TIs shown in the proper locations?   _____
12. Product analyzers required and shown?   _____
13. Vent control circuit and its destination indicated?   _____
14. Thermosiphon reboiler in specific elevation to   _____
    operate properly?
15. All above items reviewed with the instrumentation group?   _____
16. Control systems engineers reviewed the control scheme?   _____

8.10.13 P&IDs for heat exchangers

8.10.13.1 Design data

1. Equipment tag includes the following information?   _____
   a. Equipment number, Service   _____
   b. Heat Duty at design conditions   _____
   c. Surface area   _____
   d. Type (shell and tube, etc.)   _____
   e. Fan "BHP" for air coolers   _____
   f. Design pressure and temperatures   _____
      (for shell and tubes)
   g. Insulation requirement   _____
   h. Materials of Construction (for shell and tubes)   _____
   i. Size (e.g., 41" × 240")   _____
   j. TEMA Designation (e.g., AEU)   _____

8.10.13.2 Process and safety
1. When out of operation, all exchangers completely ventable and drainable?   _____
2. Appropriate information concerning design pressure/temperature, metallurgy, heat duty insulation shown on P&IDs?   _____
3. Heat exchanger location plot been approved?   _____
4. Shell and tube side protected for overpressure scenario?   _____
5. Appropriate utilities shown?   _____

8.10.14 P&IDs for furnaces or heaters

8.10.14.1 Design data
1. Equipment tag to include the following information?
   a. Equipment number, service   _____
   b. Heat absorbed at design conditions   _____
   c. Type (e.g., cabin, cylindrical)   _____
   d. Design pressure and temperature   _____
   e. Materials of construction   _____

8.10.14.2 Process and safety
1. Damper controls shown?   _____
2. Dampers provided with a locally mounted, emergency manual operator?   _____
3. Heater passes shown with economizer, preheater, etc.?   _____
4. Soot blowers required and shown?   _____
5. Snuffing steam required and shown?   _____
6. Heater controls comply with required design standards?   _____
7. Fuel gas knockout pots shown with drains?   _____
8. Multi-fuel controls provided?   _____
9. Stack sampling ports provided with stack instrumentation?   _____
10. Process stream instruments provided?   _____
11. Pilot gas system shown?   _____
12. Balancing valves shown on combustion air and fuel lines for the heaters?   _____
13. Forced draft and induced draft fans shown correctly with adequate controls and instrumentation?   _____
14. Air preheaters shown correctly with proper instrumentation?   _____
15. Double block and bleeds required and shown?   _____
16. Controls meet local, state, and federal requirements?   _____

17. Proof of closure switches provided on fuel valves? _____
18. Proper fuel interlocks (e.g., low fuel pressure, high fuel _____
    pressure, low-fire start, etc.) shown and approved?
19. Atomizing steam or air provided for heavy oil firing? _____
20. Viscosity control required for heavy fuel oil? _____
21. Manual fuel shut-off valves and utility shut-off valves _____
    provided at a safe distance (50 ft.) from heater?
22. Process flow cut-off switch provided? _____
23. High stack and high process temperature _____
    alarm provided?
24. Bypass valves around automatic fuel control valves car-sealed _____
    closed, painted red, and equipped with a warning sign?
25. Relief valves provided and properly sized? _____
26. Fuel gas lines properly drained and venting provided? _____

8.10.15 P&IDs for storage tanks

8.10.15.1 Design data
1. Tag number and Service _____
2. Size (Diameter and Height) _____
3. Capacity (barrels) _____
4. Metallurgy _____
5. Pressure, vacuum, temperature design conditions _____

8.10.15.2 Process and safety
1. Pressure and vacuum set points for all relieving devices _____
2. The main vent header in a closed vent system must show the _____
   direction of slope and note indicating "no pockets."
3. Relief devices having inlet or discharge piping _____
   have bird screens to prevent plugging
4. Atmospheric discharge piping from a relief device must indicate _____
   the elevation of the discharge point relative to the personnel
   platform near the relief device.
5. For agitated tanks, the low-level shutdown shown for _____
   protection of the agitator?
6. Tank bottom slope to low point for drainage? _____
7. The storage tanks shown correctly? _____

8. Pressure control system required/provided for:
   a. Environmental?   _____
   b. Fire (NFPA)?   _____
   c. Product quality?   _____
9. Floating roof required/provided for environmental   _____
   concerns with:
   a. Top foam/fire connections?   _____
   b. Roof water drain off?   _____
10. Fire exposure relief protection provided?
    a. Emergency manway cover   _____
    b. Weak seam   _____
11. Inert nitrogen gas required for pressure control?
    a. Two stage nitrogen let down provided?   _____
12. Flame arrestor required/provided per NFPA?   _____
13. Top and side manways provided/required?   _____
14. Heaters or cooling coils or jackets required/   _____
    provided?
15. Utility connections shown?   _____
16. Agitation required/provided?   _____
    a. Top entering for small tanks   _____
    b. Side entering for large tanks   _____
    c. Pump-around through an eductor/mixer   _____
17. Emission control required/provided?   _____
    a. Vent condenser   _____
    b. Scrubber/carbon beds   _____
    c. Incinerator/fan/blower may be required to   _____
       provide pressure
18. Overflow line provided/required?   _____
19. Overfilling protection provided/required?   _____
20. Double bottom or leak detection system required?   _____
21. Auxiliary nozzles required/provided?   _____
    a. Water drain off   _____
    b. Sump drain   _____
    c. Sample connections – possible multiple on sides   _____
    d. Temperature   _____
    e. Side drain off – possible multiple   _____
    f. Dip hatch   _____
22. Pressure storage shown correctly?   _____
23. Safety valve provided/required?   _____
    a. Blow large fire loads to air   _____
    b. 15 min holdup above High Level Alarm   _____
       if discharge to air

24. Double blocks on drain lines if LPG? _____
25. Pumps outside dike if flammable material? _____
26. Automatic block valve at vessel inlet/outlet _____
    provided/required?
    a. Toxic material _____
    b. Flammable material _____
27. Block valves required/provided on all connections _____
    below the liquid level?

8.10.16 P&IDs for boilers

1. Boilers shown correctly? _____
2. Fuel gas firing controls per NFPA standards? _____
    a. Automatic isolation of all fuels _____
    b. Flame failure shutdown _____
    c. Fuel pressure monitoring _____
    d. Combustion air flow monitoring _____
    e. Pilot flame system _____
    f. Ignitor/light-off system _____
3. Dual safety valves required/provided? _____
4. Blowdown to a safe location? _____
    a. Continuous on flow control or conductivity control _____
    b. Manual for shock blowdown _____
    c. Multiple blowdown taps required _____
5. ASME and NFPA codes complied? _____

8.10.17 P&IDs for lube oil plans

1. Lube oil systems shown correctly? _____
2. Main auxiliary pumps shown/required? _____
3. Lube oil cooler shown/required? _____
4. Lube oil filter(s) shown/required? _____
5. The control show? _____
    a. Oil pressure _____
    b. Oil temperature _____
    c. Oil flow _____
    d. Auxiliary Oil Pump (AOP) start-up _____
    e. Shutdown system _____
6. Lube oil heater required/shown? _____

8.10.18 P&IDs for flares

1. Flare systems shown correctly?
2. Inlet knockout drum required/shown for separation
    of liquid and vapor?
3. Control system shown?
    a. Pilot system
    b. Ignitor system
    c. Steam/air assist for smokeless operation
    d. Assist fuel for noncombustible mixtures
    e. Flare seal fluid level
    f. Freeze protection
4. Flare fluid seal pumps required/shown?
5. Inlet lines adequate for auto refrigeration
    temperatures?

Chapter 9
Safety systems for the plant

Safety systems for chemical or refining plants are critical and are discussed in this chapter. As the process unit deals with hazardous chemicals, the process can create an unsafe situation if safety provisions are not provided. If the safety systems are not provided, personnel working near the plant and the plant itself are at risk. There are different safety systems for a given plant. For example, a fire water system is used to suppress the fire. During an external file scenario, the deluge safety system cools the vessel and piping surface. Safety showers and eyewash stations are used by personnel exposed to chemicals or hazardous materials. Flare knockout pot and seal pot are used to remove unwanted liquids that can give rise to safety concerns if burned at the flare. A flare network is used to collect different flares from different units; eventually, the collected flare material is burnt at the flare.

9.1 Flare

The flare system consists of a flare stack, a flare back prevention section, a flare tip, ignition mechanism, a steam injection manifold, and a fuel gas burner management system. The flare stack height is typically greater than 80 ft. and depends on the capacity of the flare stack and the radiation it creates during the burning of the flammable material. Steam is used to clean up the flare tip nozzles, cool the flare tip, and control the flame pattern. Similar to the boiler, the flare system also has a fuel gas burner management system. In addition to that, a pilot gas is always in operation, and the flare is always burning at its minimum capacity.

9.2 Flare network piping

Different relief headers in the process plant are connected to a common flare header in the unit. All such flare headers from different units are tied into a plantwide flare header network. The flare piping system can see gas, liquid, or mixed phase. It is important to note that all the headers and sub-headers are designed so that the liquid contents in the flare line are free-drained to the flare knockout pot. The larger plantwide flare header network is big enough to handle the global scenarios from individual units and plantwide global scenarios such as instrument air failure. The flare network involves pipe connections, valves on the unit's flare sub-headers or flare headers, and a common flare meter. A flare meter just before the flare tip measures the actual hydrocarbons to be burned in the flare over time, and trends are observed. Figure 9.1 shows a typical flare network system. The materials, which are very heavy at normal atmospheric temperatures and

https://doi.org/10.1515/9781501519864-009

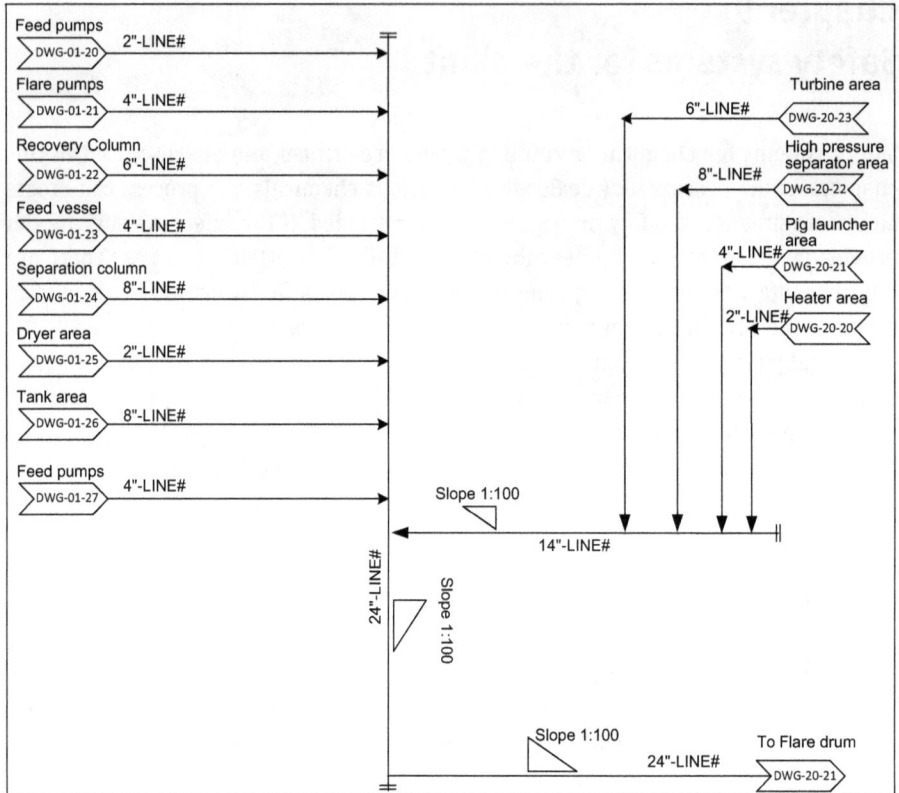

Figure 9.1: Flare network system.

pressures, would need external heat to keep them in the liquid state. For such fluid services, steam or electrical tracing is usually provided around the flare piping from the source of the heavy material relief to the flare knockout pot. Electrical tracing is most used in industries as it is advantageous compared to steam tracing. Steam tracing would need additional infrastructure such as additional low-pressure steam, steam piping, condensed collection piping, steam traps, etc.

9.3 Flare knockout pot and seal system

As the flare network sees liquid or mixed or gas phase, removing liquid before the flare tip is important as liquid hydrocarbon carrying over to the flare tip can create a safety hazard. To overcome such problem, a flare knockout pot is installed just before the flare stack and designed for the maximum liquid volume possible from any unit or worst-case relief scenario within the plant. Once the liquid is collected in the flare knockout pot, a set of pumps transport the collected oil or chemicals back to the processing facility, depending on the level of the flare knockout pot. If heavy material is expected from the flare network, it is recommended to steam or electrically trace the entire flare knockout pot to ensure that the material received does not solidify.

As the flare material is hazardous and flames from the flare stack can backflow into the relief headers and networks, it is critical to create a seal between the relief network and the flare stack. There are different mechanisms to create a liquid seal. The most commonly used seal mechanism is by adding an overflow weir plate. The relief material comes from the relief network and is discharged into the liquid side of the seal chamber through a dip pipe. The separated gas travels upward towards the flare stack for burning. A 2"-line of the utility water is connected to the seal chamber and is always open, making up the water in the seal chamber. The overflow of water gets into the other side of the seal chamber. The design of the seal leg and the height selected for the seal leg help maintain a constant level on the other side of the seal chamber. This way, flammable gases in the flare stack are positively separated from the relief material from the flare network.

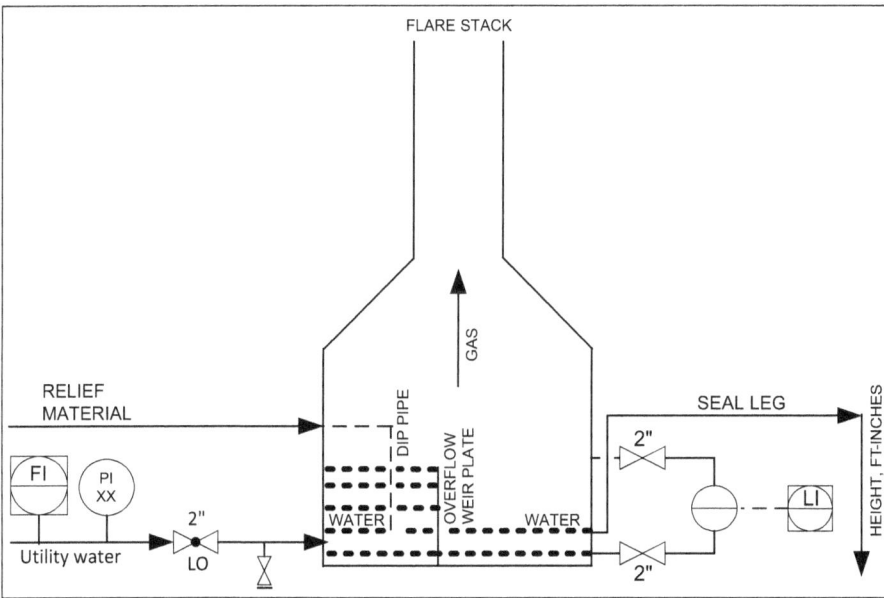

Figure 9.2: Flare seal system.

9.4 Fire water system and network

The fire water system involves a fire water tank, pump, network system, and fire water delivery safety equipment. The entire fire water system is designed as per National Fire Protection Association (NFPA) code. The fire water tank is designed to handle a worst-case fire scenario within the plant. Typically, the clarified water or utility water is used as makeup water for the fire water tank. A control valve on the feed line ensures that the tank is always liquid full. At least one of the fire water pumps needs to be diesel-driven. The fire water pump's dedicated relief valve outlet handles overpressure scenarios such as a blocked outlet. The relief from the relief valve is routed back to the tank. The fire water tank has an overflow protection pipe that protects the tank from overfill scenarios.

The fire water network distributes the water to different headers and different units. A detailed network and pumps need to be run to ensure adequate pressures are reached at all the firewater monitors and hose stations. When adequate pressures are not reached based on the modeling, the network header size needs to be increased. A firewater network can also supply water to the fire water sprinkler system. Since the fire water header is always pressurized, the sprinklers may need maintenance or replacement over time, due to the impact of water.

9.5 Deluge safety and fire and gas monitoring system

Deluge systems are commonly used for chemical plants or hazardous chemical plants where the entire surface of the plant needs to be wetted with water very quickly to extinguish the fire. A deluge system is very similar to the fire water sprinkler system. The primary advantage of a deluge system is that the sprinkler piping is not pressurized with water and does not contain any water during normal scenarios.

The trigger for activating the deluge safety system is based on fire and gas monitoring. When a fire is detected at the monitoring station, a set of pumps provides utility water to the deluge system.

9.6 Safety shower and eyewash stations

Safety showers and eyewash stations are used to clean the hazardous chemicals immediately after exposing a person to the hazardous chemicals. Every unit has multiple safety shower and eyewash stations. The bigger the plant size, the higher the number of safety showers and eyewash stations. Safety shower and eyewash stations are part of a common safety system.

Potable water is not processed or treated in the processing unit. The water comes straight from the local municipality. Potable water is primarily used in the safety shower and eyewash stations. Depending on the plant's location, the safety shower piping is usually electrically traced to avoid freezing due to winter conditions.

Chapter 10
Drafting, checking, master markup (MMU), and revisions

This chapter discusses details around drafting, checking, and revising the P&IDs. P&IDs are first marked by engineers and then drafted by the drafting team. Later, the P&IDs are checked for any mistakes. Depending on the phase of the project, the P&ID drawings for the project are issued with a certain revision title. Drafting and checking cycles are important to ensure that the drawing's quality is maintained. The revisions for the drawings are required to distinguish the earlier phase from the current phase. The customer must approve all the marks in Phase-4 through the master markup (MMU) procedure.

10.1 Drafting

When the chemical process engineer completes the marks on Bluebeam, the lead process engineer will notify a team with drafting expertise. There are multiple phases in any project. Phase one, the scope planning, does not involve any drafting of the P&ID. In phase one, a rough sketch of the process concept or a preliminary sketch that is prepared using Adobe software is adequate. Phase two is the definition phase; in this phase, the scope of the process is finalized. The P&IDs are redlined in the Bluebeam but are not drafted in this phase. In this phase, other disciplines are not involved, so the high-quality drafted P&IDs are not required, and the P&IDs are called issued for an estimate (IFE). The next phase is phase three, which is scope development. All other disciplines, such as mechanical piping, civil, electrical, instrumentation, etc., are involved in developing the scope. There is great pressure on the chemical process engineer to develop P&IDs as quickly as possible. This phase is also the first time the drafting engineers are involved in drafting the P&IDs. In the last phase, which is the detailed design, the P&IDs are further developed based on several factors such as vendor information, the relief valve calculations, review meeting between a customer and the EPC contractor, etc. The developed P&IDs are further drafted and later issued to the customer.

Chemical engineers or process engineers, including other disciplines, are responsible for preparing P&IDs for the project. Initially, the process engineer completes the marks on the P&IDs and notifies other disciplines to add their marks. There is always communication among other disciplines and processes through emails or phone or personally in the office. Once all the marks are added to the P&IDs, the process engineer prepares a list of drawings to be drafted, and a drafting coordinator is notified. The drafting coordinator looks at the marks on the P&ID and gets back to the process

https://doi.org/10.1515/9781501519864-010

engineers and other disciplines if there are any questions. If there are no questions, the drafting coordinator sends the drawings to the drafting team. Once the drafting is complete, the drafting coordinator is notified by the drafting team. Later, the process engineers and other engineers are given the drafted copy of the P&IDs and requested to back-check their marks. Once the engineers check the drafted copy, if the draft marks are acceptable, the process engineer signs off on the drawing and the drafted copy is issued through document control. If the marks are unacceptable, the engineers will notify the coordinator, and the corrections are drafted again, and the cycle repeats. The entire drafting process is shown in Figure 10.1.

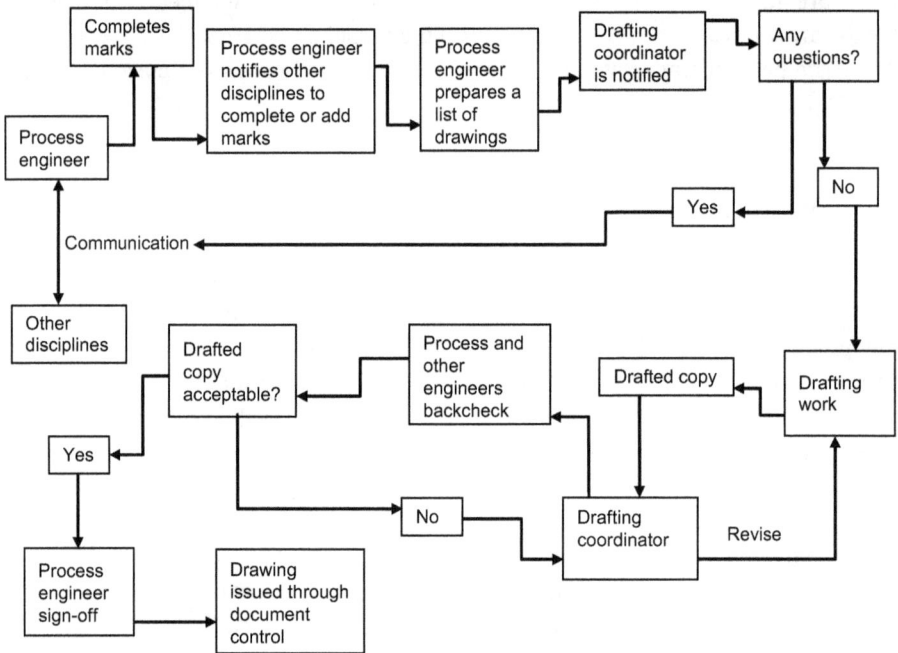

Figure 10.1: Drafting work process used in general.

10.2 Checking

The drafting team does the job of drafting the P&IDs. As a first pass, the drafting team lead needs to ensure the drafter has worked on all the comments given on the P&IDs. Once accepted, the drafting team lead notifies the drafting coordinator about completing the work. The drafting coordinator further notifies the process and other discipline engineers and requests their feedback. If there are further comments from the back-checking, the comments are given to the drafting coordinator, who works with the drafting team to complete those corrections. This cycle continues until no further comments from the process and other discipline checkers. Every time process and other disciplines back-check a copy of P&IDs, they are required to add their initials and date, as shown in Table 10.1.

Table 10.1: Check the record box sample for a document.

Drawing Back-Check Status		
Date of drafting completion: _____		
Action	Initials	Date
Drafting team lead back-checked	XX	XX
Drafting coordinator (initiate checking process)	XX	XX
Back-checker 1 (process engineer)	XX	XX
Back-checker 2 (other discipline engineer)	XX	XX
Back-checker 3 (other discipline engineer)	XX	XX
Drafting coordinator (corrections to drafting team)	XX	XX

10.3 MMU procedure

MMU is a management of change or master markup process. MMU practice is usually used in the detailed design where many changes happen because of vendor data and scope development. The procedure involves systematic steps to mark up items on the P&IDs and get them approved by the customer. The MMU procedure helps avoid marks incorporated into the design without an official review by the customer.

In the MMU meeting, each discipline explains its marks and discusses the reasoning behind the marks with the customers. The MMU item is approved if the customer agrees with the marks. Figure 10.2(a) shows the mark added by a discipline color-coded in dark black, which is a bypass piping and a valve across the control valve, and Figure 10.2(b) shows the marks approved by the customer. Note that Figure 10.2(a) does not show customer approval in the triangle for MMU-02, and Figure 10.2(b) shows customer approval in the triangle for MMU-02. Once the customer approves the marks, the drawing can be sent for drafting purposes.

Also, during the approval process, the customer decides whether a particular MMU item needs to be studied for operational hazard study (HAZOP). Table 10.2 shows an example of MMU addition with a change description and a reason for adding a change. Table 10.3 shows an example of MMU approval and yes/no options for HAZOP.

Changes to the MMU marks can be simple or complex. Minor changes like adding a tag number to the instrument or piping or spelling corrections do not need the customer's approval. On the other hand, the changes, such as adding a pump or a valve, which changes the flow or chemistry of the process, needs customer approval before the change can be incorporated into the design.

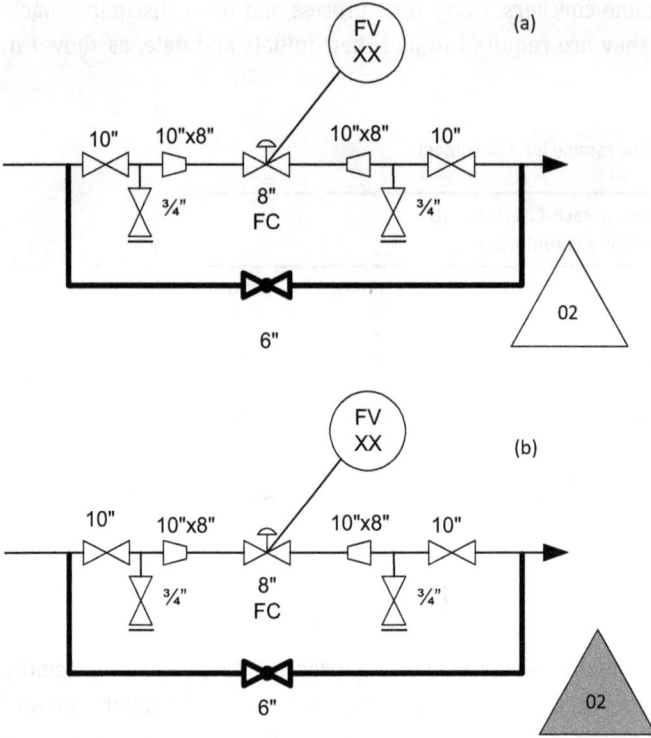

Figure 10.2: An example of MMU approval by the customer.

Table 10.2: An example of MMU addition with change description and a reason for adding a change.

P&ID number	Author	MMU number	Change description	Reason for change
DG-01-25	Avinash Karre	01	Added a bypass around control valve FV-01	Bypass recommended as per the design

Table 10.3: An example of an MMU log and associated MMU item.

P&ID number	Author	Creation date	HAZOP required, yes/no	HAZOP completed, yes/no
DG-01-25	Avinash Karre	02/08/2023	Yes	No

10.4 Revisions and project issue

When an operating company gives the project to an engineering contractor, it is usually in phase 1. In phase 1, the focus is usually on the development of the process with the help of PFD, and no actual work is done on P&IDs. In phase 2, an engineering contractor develops the P&IDs to show the preliminary scope, and one cannot use this information to build a plant. P&IDs are issued as "Issued for an estimate (IFE)" in phase 2. Phase 3 is when the actual development of P&IDs takes place. The development of P&IDs is required based on the development of equipment design and feedback from many equipment manufacturers. There are four P&ID issues in phase 3, namely, "Issued for review (IFR), Issued for approval (IFA), Issued for HAZOP (IFH), and Issued for design (IFD)." Phase 4 involves finalizing the equipment manufacturer and preparing for isometric drawings, and issuing "Issued for construction (IFC)" P&IDs. Several revisions (such as Issued for construction – Revised (IFC-R)) are often seen mainly because of changes in the pipe routing and equipment design feedback from the manufacturers. Table 10.4 summarizes all the revision cycles in different phases of a project.

Table 10.4: Revision cycles for P&IDs (with tie-in/demo P&IDs included).

Phase of the project	Revision title	Description	Revision numbers
2	IFE	Issued for estimate purpose	A,B,C, etc. (Internal issue)
3	IFR	Drawings issued for line-by-line review	0, 1, 2, etc. (Issued to customer)
3	IFH	Drawings issued for the HAZOP review meeting	0, 1, 2, etc. (Issued to customer)
3	IFA	Drawings issued after the HAZOP comments are picked up	0, 1, 2, etc. (Issued to customer)
3	IFD	Once design progress is complete towards the end of Phase 3, drawings issued	0, 1, 2, etc. (Issued to customer)
4	IFC	Issued for construction	0 (Issued to customer)
4	IFC-R	Issued for construction – Revised	1 (Issued to customer)

Chapter 11
P&ID systems

P&ID systems consist of different unit operations in a processing plant. All these different unit operations are commonly used in all industries. This chapter provides a summary of the operation of the most used P&ID systems and details to develop them.

11.1 Control valve stations

Figures 11.1(a) and (b) show the typical arrangement of control valve stations. Each control station functions based on a certain input variable. The actual control valve body is at the center of the control valve station. Upstream and downstream of the control station, a block valve can be used to isolate the control valve in case of maintenance activities. During such maintenance activities and when the control valve is out of service, a bypass valve and a line can be utilized. The bypass valves are usually globe-type, providing the necessary flow control function. The isolation block valves are typically gate valves to provide the lowest pressure drop in the process system. Based on the process parameters, such as pressure drop and flow rate, the size of the control valve is determined, which is typically smaller than the actual line size. For example, in Figure 11.1(a), the actual line size is 18 inches, while the control valve's size is 12 inches. The size of the bypass valve depends on the control valve's rated Cv (characteristics of a valve). And usually, the bypass globe valve is selected to have a very similar Cv compared to the control valve, to account for any possible relief scenarios. Since the control valve is smaller than the actual line size, the control valve always has upstream and downstream pipe reducers, as shown in Figure 11.1(a). When the control station is removed for maintenance, it is necessary to drain the pressure and the flow in the section between the two isolation block valves, to safely remove the control station. To do that, ¾" drain valves are provided on the upstream and downstream sections of the control valve, as shown in Figures 11.1(a) and (b). The actual type of the control valve depends on the service of the fluid and the available pressure drop across the system. For example, if a robust flow control is needed, then a globe-type control valve is always required. For situations where a robust flow control is not necessary, a butterfly valve type can be used. Figure 11.1(a) shows a control valve of the globe type, and Figure 11.1(b) shows a control valve of the butterfly valve type.

Usually, there is a pressure change across a control valve, and it may be necessary to add a piping specification break right after the control valve or downstream of the second isolation block valve to lower the cost of the process piping system. As the cost of equipment and piping increases exponentially with increase in pressure rating and pipe class, it is always recommended to add such piping specification breaks, where needed. Figure 11.1(a) shows such breaks. Also, each control valve is

https://doi.org/10.1515/9781501519864-011

designed for a fixed fail-safe position, which can be either fail-close or fail-open. The arrangement of the instrument tubing on the diaphragm of the control valve can achieve the fail-safe position of the control valve. For a fail-closed valve, the instrument is supplied from the below the diaphragm, and when the instrument air fails, the control valve goes to its closed position as a failsafe mechanism.

The failsafe mechanism of the control valve should always be shown below the control valve, where the control valve size is shown. For fail-closed valves, "FC" can be shown, and for fail-open valves, "FO" can be shown. On the other hand, for a fail-open valve, the instrument is supplied from the top of the diaphragm, and when the instrument air fails, the control valve is opened fully by the spring as a fail-safe mechanism. An example of a failed-close valve is shown in Figures 11.1(a) and (b).

Note that most processing plant control stations are at grade levels. For a complex plant with multiple levels to support columns, reactors, and pressure vessels, control valves are located at the floor level on each floor or at grade. Control stations are

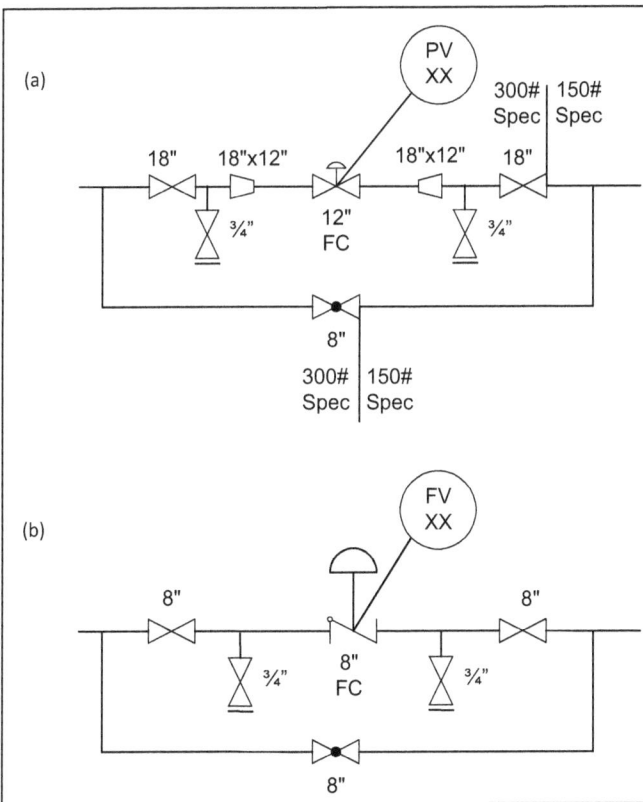

Figure 11.1: (a) Typical control station with globe-type of valve; (b) typical control station with butterfly-type valve.

located at grade or at a reasonably accessible location so that maintenance people can carry out the maintenance activities on the valves.

Line numbers are required for all pipes when making the P&IDs. Line numbers in Section 11 of this book are not provided, for simplicity and to avoid clutter in the figures, which are discussed in subsequent sections.

11.2 Steam letdown stations and desuperheater

Steam letdown stations are a little different than other control stations. This is because there are different levels of steam required for a process plant. The different steam levels are 600 PSIG, 250 PSIG, and 50 PSIG. A boiler makes steam at 630 PSIG, which is distributed to different units through steam letdown stations. Different units have different temperature requirements for heating, and the demand for different steam levels could vary, depending on the process unit usage.

Moreover, all levels of steam need to have some level of superheat in them to avoid excessive condensation in the piping network. Also, desuperheating or removing the excess temperature from the superheated steam is done to meet the process operating temperature requirements. Figure 11.2 shows the steps involved in the steam letdown stations and in the desuperheating operation.

In the first step, 600 PSIG steam is letdown to 250 PSIG steam. The produced 250 PSIG steam is distributed to 250 PSIG users and also let down for 50 PSIG users. When the 50 PSI steam is produced, it has a high degree of superheat, which needs to be removed to avoid exceeding the design requirements of the 50 PSIG steam users in the process unit. The high degree of superheat is removed by adding a desuperheater. A boiler feed water or recovered condensate from the plant is typically utilized to desuperheat the superheated 50 PSI steam. As the water at lower temperatures is added to the desuperheater, the superheated 50 PSIG steam temperature is lowered to meet the user's requirement. A control valve is added to the boiler feed water line that modulates the water flow rate, depending on the temperature requirement of the 50 PSIG steam system.

As the 600 PSIG steam is letdown to 250 PSIG steam, there is always a safety risk that the 250 PSIG steam network and its users could potentially see a 600 PSIG steam pressure when the 600 to 250 PSIG steam letdown fails due to a malfunction. In order to protect the 250 PSIG steam system from such an overpressure scenario, a relief valve is added and set reasonably well above the operating pressure of the 250 PSIG steam system. This relief valve relieves the overpressure to the atmosphere at a safe location. Similarly, when the 250 to 50 PSIG steam let down fails due to a malfunction, the 50 PSIG steam network and its users could potentially see the 250 PSIG steam pressure, which is unacceptable. For such reasons, a safety relief valve is added downstream of the letdown, and set reasonably above the operating pressure of the 50 PSIG steam system. This relief valve also relives to the atmosphere at a safe location.

When steam travels through pipelines, heat is lost as the steam condenses into a water phase. If the condensed steam is left inside the piping, it could create a water

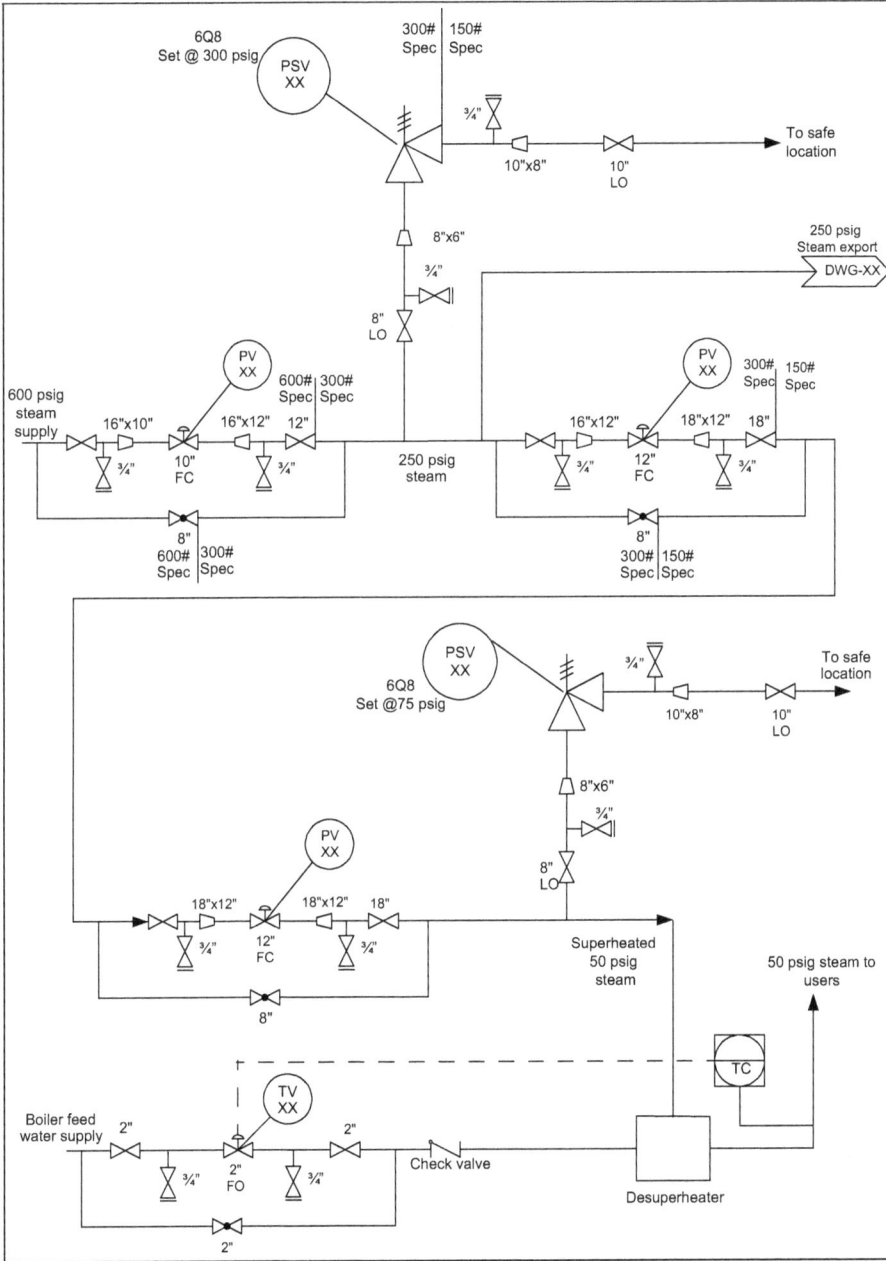

Figure 11.2: Steam letdown stations and desuperheater arrangement.

hammering effect, and it is necessary to remove the water from the piping. Also, it is important to remove the condensate before the steam enters the control station. As an example, in Figure 11.2, since the steam is superheated, the steam trap may not be necessary as the likelihood of condensate formation is low. But for situations where saturated steam exists, steam traps are always required to avoid condensation inside the control valve, which can cause operational problems of the control valves. The entrained condensate can reduce the capacity of the steam control valve, which can lead to poor performance of the valve.

11.3 Steam traps and types

A steam trap is a specialty piping component that traps the steam and removes the condensate from the steam line. Float-type steam traps are very commonly used in the industry. The steam trap has a float inside and works on a density difference principle. The density of the condensate is higher than the density of the steam. As the condensate builds up in the steam trap, the float inside moves in the upward direction, allowing the condensate to escape. Once all the condensate is removed, the float comes back to its original position, trapping the steam.

Figure 11.3(a) shows details of a simple steam trap where the condensate produced is collected in the drain hub, and there is no further recovery of the condensate. This type of arrangement may be acceptable where a small quantity of steam is used. Figure 11.3(b) shows trap details where the condensate produced is collected and recovered through the condensate recovery system. Figures 11.3(a) and (b) are the most used trap details for removing condensate from exchangers. Figure 11.3(c) shows a trap located on the 50 PSIG steam line. As most of the time, the 50 PSIG steam is available slightly above the saturated point; more traps are needed for every 250 feet, depending on the heat losses from the system. Drip legs, consisting of 6" pipes, collect the condensate from the 50 PSIG steam pipe, and the trap removes the as produced condensate. Note that steam traps are made of a higher piping class to avoid overpressure, and a piping specification break is added downstream of the trap. Isolation valves can be added to the inlet and outlet of the steam trap to carry out routine maintenance.

11.4 Pressure safety valve: liquid service

Pressure safety valves are safety devices that protect the equipment or the piping system from overpressure. There could be several relief scenarios for a given system. The relief valve design can have liquid, vapor, or two-phase relief. Liquids that are flammable and have hazardous properties cannot be released into the atmosphere, and they cannot be sent directly to the flare for burning, due to safety reasons. All

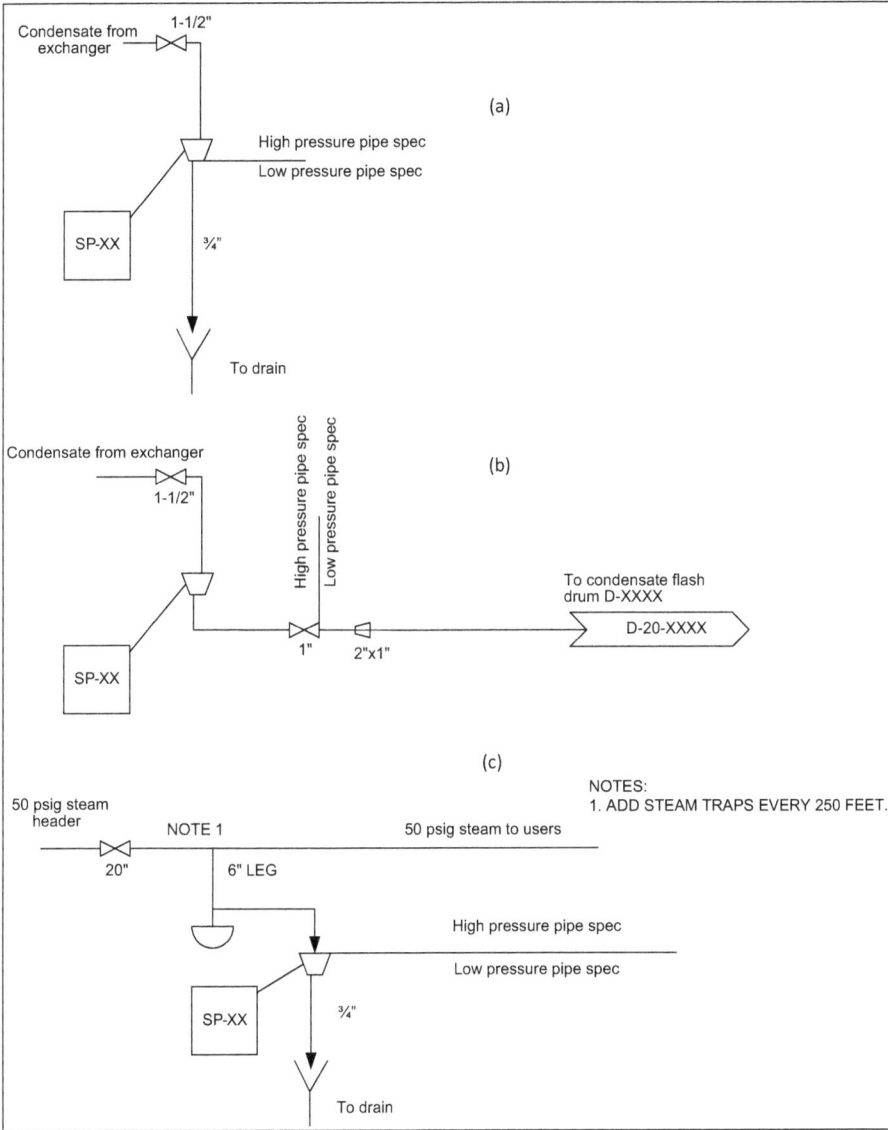

Figure 11.3: Steam traps and types.

such flammable and hazardous liquids are routed to the flare drum or a knockout pot, where liquids are separated from gases. The separated liquid is pumped out using a set of pumps, and the separated gases are sent to the flare for burning. Relief valves that are designed for non-hazardous and non-flammable services such as water or boiler feed water service can be routed to a safe location near the grade.

Figures 11.4(a) and (b) show examples of such relief valves relieving only liquid or two-phase services. Figure 11.4(a) shows the relief valve that relieves hazardous chemicals or flammable material into the knockout pot through the flare header. Figure 11.4(b) Shows the relief valve that relieves non-hazardous materials such as water. Figure 11.4(a) shows a relief valve with inlet and outlet isolation valves. These isolation valves are either car-sealed open or locked open to avoid accidental closure of these valves as these valves

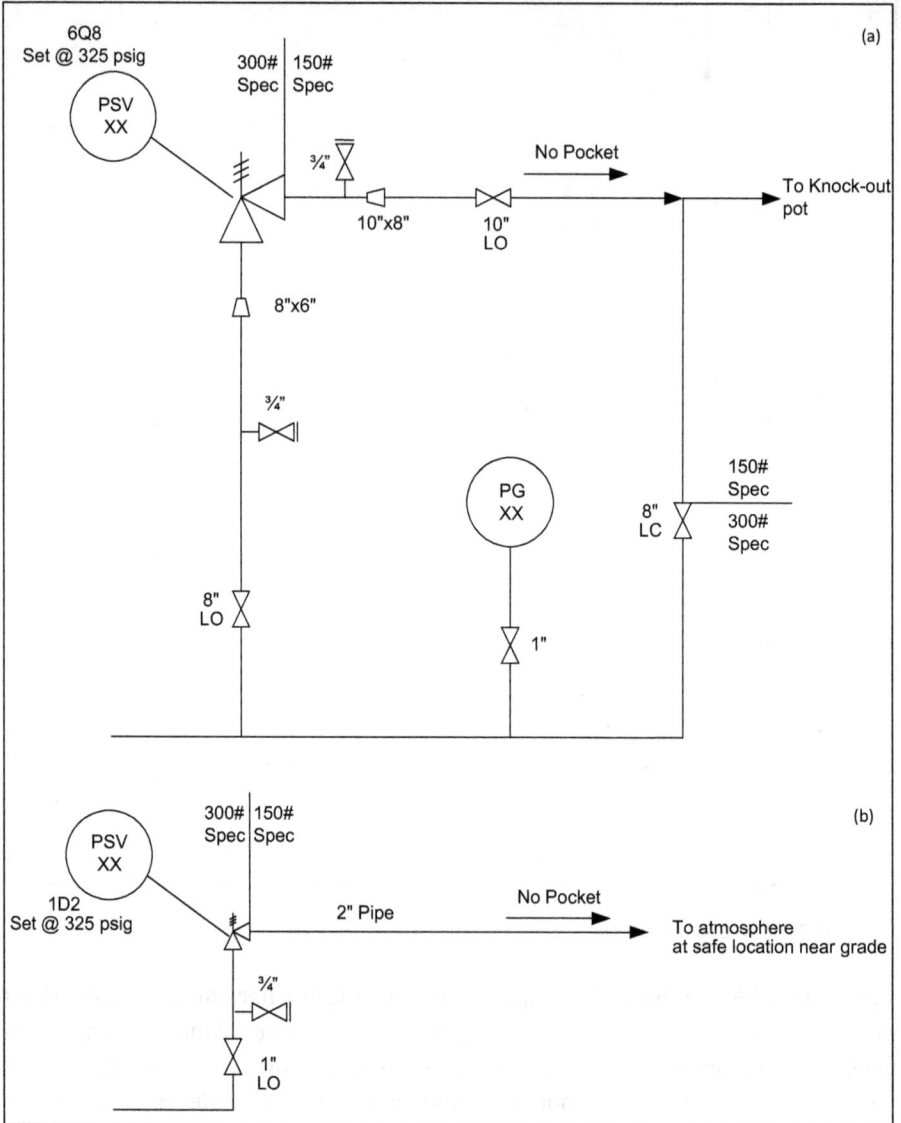

Figure 11.4: Pressure safety valve of liquid service.

provide a relief path through the relief valve. A bypass is provided with a valve and used when routine or safety maintenance is needed on the relief valve. The bypass valve is locked closed or car-sealed closed in normal operation. All these block valves are full-port. During such maintenance, pressure needs to be removed from the section between the block valves and the safety valve; for such scenarios, ¾" bleed valves are provided. Note that the safety valves come in different sizes. Inlet and outlet piping sizes depend on the pressure drop criteria for a relief valve. Depending on the relief valve's size and inlet and outlet piping, pipe reducers are commonly needed upstream and downstream of the relief valve. Pipe specification breaks are needed just downstream of the relief valve and downstream of the bypass valve as the relief valve downstream system is a low-pressure system and could have different piping metallurgy compared to the process where the relief valve is located. It is a common practice in the industry to also add a pressure gauge on the inlet piping of the relief valve to check the pressure at the inlet of the relief valve or process system as part of routine field checks carried out by process operators.

Relief valves are identified by a tag number placed inside a bubble. Outside the bubble, information such as the set pressure of the relief valve and the size of the relief valve are mentioned. Designing the inlet and outlet piping of the relief valve is critical, and for safety reasons, a pocket is not allowed in the pipe. If a pocket is present in the piping, liquid can accumulate over time and act as resistance when relief action is needed. It is a common practice to add a "no pocket" note on the relief valve inlet and outlet piping. In addition, the outlet piping of the relief valve is always designed to be "free draining" so that it free drains to the flare header or the flare drum. If the piping is not designed for the material to be free-drained, it will lead to the formation of an undesirable pocket. As shown in Figure 11.4(b), some relief valves do not have a bypass. Relief valves without a bypass might be acceptable for systems with spare equipment or clean services, such as soft water.

11.5 Pressure safety valve: gas or vapor service

Figure 11.5(a) shows a hazardous vapor service relief valve. Relief from the relief valve goes to the flare header. Relief from the flare header is burned in the flare. Figure 11.5(b) shows a non-hazardous vapor service relief valve such as steam. The relief from this relief valve is routed to the atmosphere at a safe location. Oftentimes, the operational relief valve needs to be taken out for an extended period of time for maintenance. In such scenarios, a bypass around the relief valve may not work as a viable solution. As an alternative, full spare relief valves are always added in parallel with the one in operation. This full spare arrangement of the relief valve makes the overall process system more robust and flexible during maintenance activities. Such an example is shown in Figure 11.5(a). For non-hazardous relief valve service, 4 mm weep holes are provided at the elbow in the discharge piping to remove any condensed liquid in the discharge piping of the relief valve. Since the relief valve outlet is open to the atmosphere, debris from the atmosphere or dust can accumulate inside the discharge piping. Birds can also create a nest in the discharge piping, creating restrictions during relief events. For such scenarios, usually bird screens or weather protection caps are installed. A specialty item represents a bird screen or a weather protection cap, shown in Figure 11.5(b).

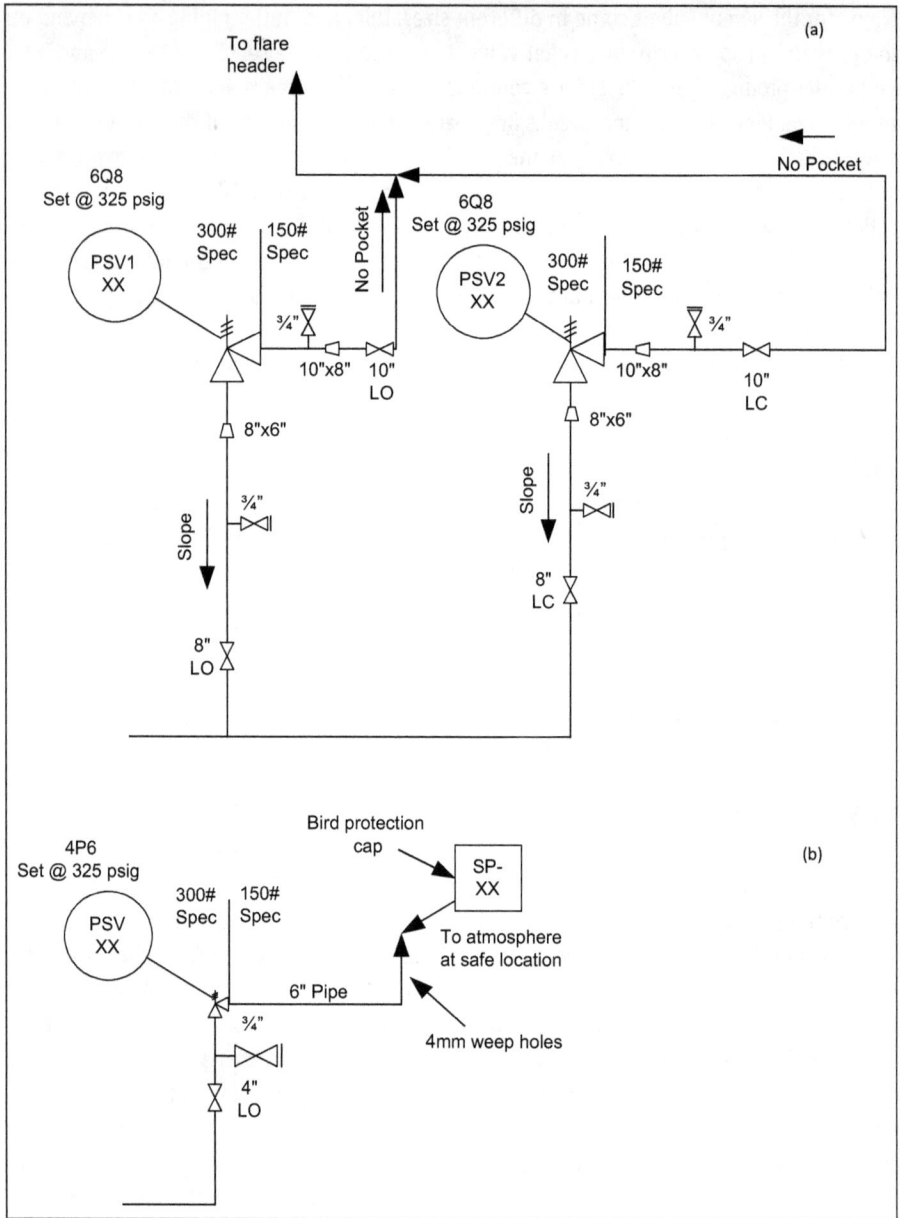

Figure 11.5: Pressure safety valve for vapor service.

11.6 Pump with restriction orifice in the minimum circulation line

Centrifugal pumps are designed with a certain minimum flow value. An operating flow above this minimum flow value is recommended for the safe operation of the pump. If the operating flow falls below the minimum recommended value for the pump, the pump may see a major problem. Pumps with a flow rate of 50 gpm or a motor size of 50 horsepower and below need minimum flow protection. For pump systems, the recirculation line from the pump discharge is routed back to the pump suction vessel, with the restriction orifice designed to pass the minimum recommended flow for the pump all the time. The minimum circulation flow line with the restriction orifice is not an ideal way to handle such minimum flow concerns, but it is economical for a pump design that has a design flow of 50 gpm or less.

Strainer in the pump suction is always added to avoid damage to the pump internals from foreign materials such as metal pieces or solids. This strainer has stainless steel internal, specifically designed for the fluid service being pumped. The mesh size of the internal of the strainer is carefully chosen to capture solids from the upstream process system. The drain valve attached to the strainer helps in removing the liquids from the strainer. Regular cleaning of the strainer is necessary as part of the preventive maintenance program.

To monitor performance, pumps are always installed with a pressure gauge instrument on their discharge. When operators go out to check the conditions of the process equipment, they always look for the discharge pressure on the pump, which is one of the performance indicators. Pressure gauges can be isolated and replaced when the pump is still running. The pressure gauge range is based on the requirement of the process.

The pump discharge system always has a check valve that prevents the backflow of high-pressure fluid from the pump discharge from getting back into the pump suction, which may be rated for low pressure. Since the suction vessel of the pump is always designed for any potential backflow from the pump, the suction piping, especially downstream of the suction-side valve and the pump suction, may not be protected when the operator closes the pump suction valve. For such scenarios, the piping between the pump suction nozzle and the pump suction valve is always rated for the same class as pump discharge, as shown in Figure 11.6.

The flow from pump discharge needs some control to be able to operate at the maximum efficiency point of the pump curve. For example, if the pump draws too much flow rate, a pump is operating at a low-frequency point, and most of the liquid could be recycled back to the suction vessel as the process does not need it. Typically, a control valve located on the pump discharge regulates the level in the pump suction tank. As the level in the vessel increases, the control valve on the pump discharge opens, which in turn activates the pump to put out more flow to take care of the increased level. This is a common logic used in the industry to send excess material to tanks.

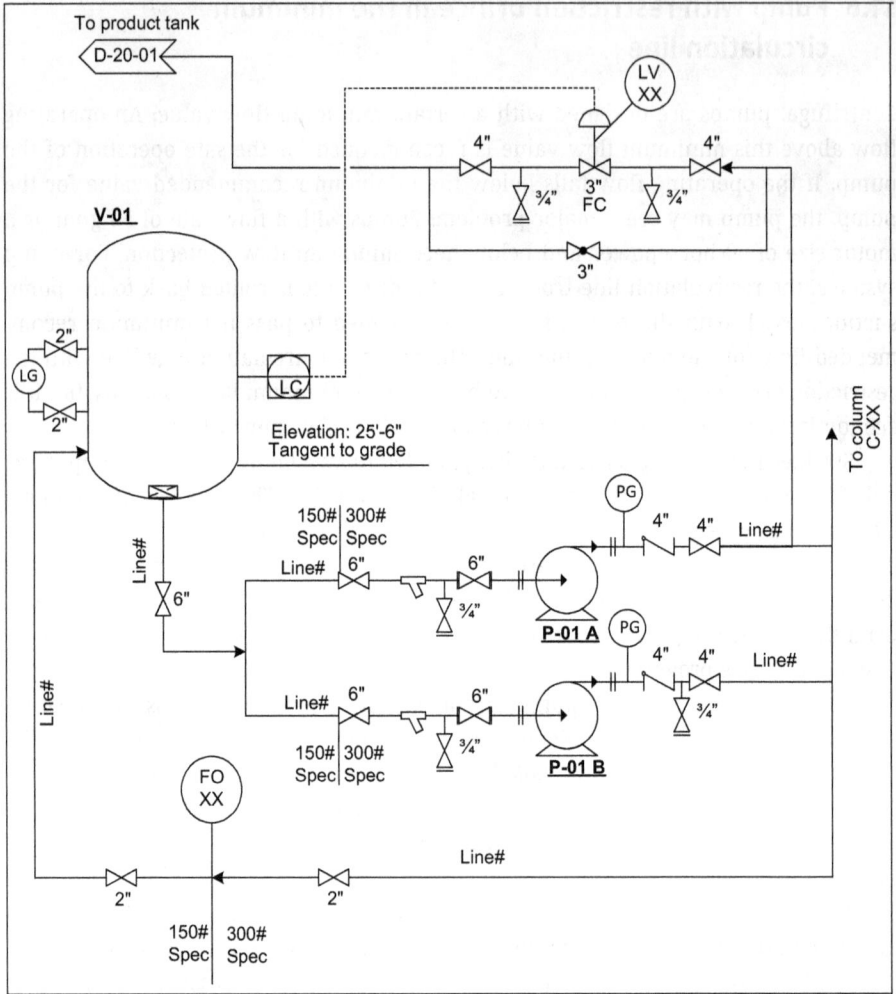

Figure 11.6: Pump with restriction orifice in the minimum circulation line.

11.7 Pump with control valve station with a minimum circulation line

Larger pumps have higher minimum flow requirements, and continuous circulation of the minimum flow cannot be allowed due to wastage of power. In such cases, it is economical to add a control station designed for a minimum flow rate of a pump. Such a system is shown in Figure 11.7.

Flowmeter is added to the pump discharge before they take off for the minimum circulation line. This meter always measures the total flow coming out of the pump.

When the pump's forward flow goes below the pump's minimum flow requirement, the recirculation line opens (through a control valve), sending more liquid back to the pump suction, and protecting the pump.

The control valve on the minimum circulation line is designed to maintain at least the pump's minimum flow requirement and is intended for the "fail-safe open (FO)" type. The control valve located on the pump discharge line controls the level in the pump suction and is always designed for fail-closed (FC) – from emptying the suction vessel of the pump, which could cause damage to the pump.

Every pump needs a certain minimum net positive suction head (NPSH), also known as NPSH-required. The available and NPSH should always be greater than the

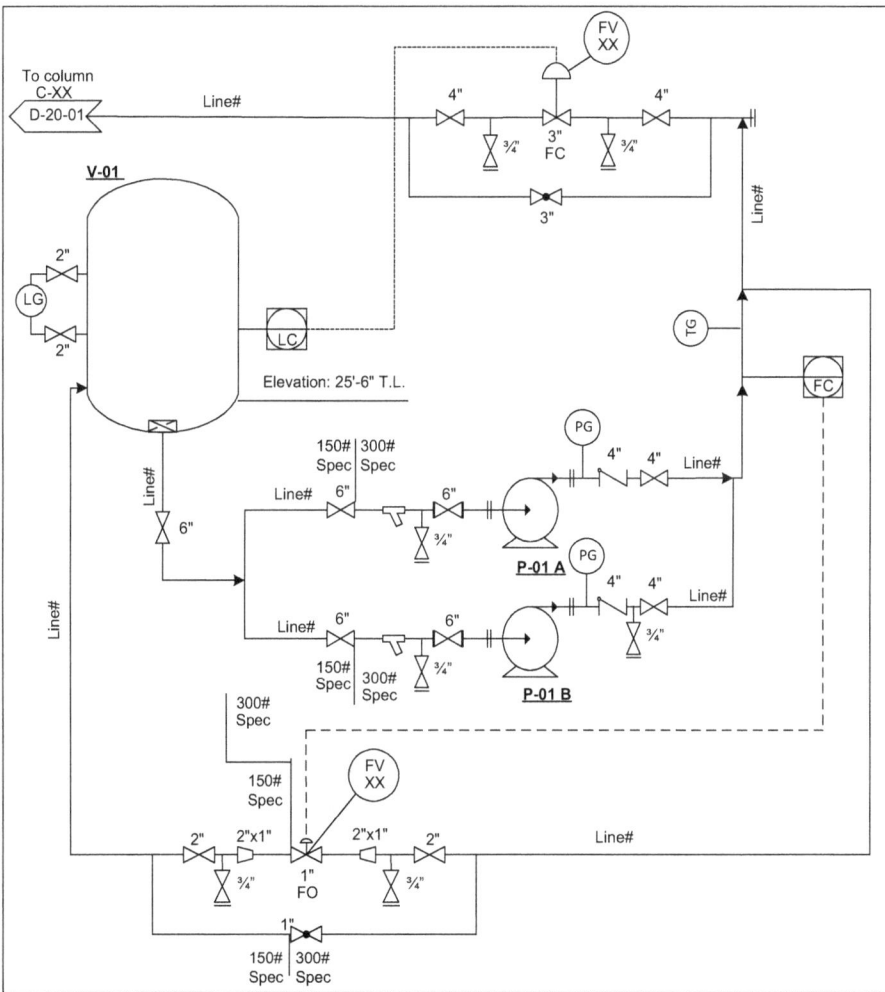

Figure 11.7: Pump with control valve station with a minimum circulation line.

NPSH-required; otherwise, the pump is likely going to cavitate, and eventually damage the pump. This is particularly important for hot liquids at their bubble point or even for low vacuum services. Liquids that are subcooled are less likely to face NPSH problems due to the low vapor pressure of the fluid. To increase NPSH, the process engineer needs to increase the height of the suction vessel and also the size the suction piping of the pump appropriately, per customer guidelines.

As noted above, the pump suction vessel is typically rated for a much lower pressure than the actual rating of the pump. It may be necessary to add a piping specification break downstream of the minimum circulation line control valve until the pumps suction vessel to match the suction vessel piping class, to save cost.

11.8 Pump drains and vents

Every pump undergoes maintenance as part of the preventive maintenance program or due to some mechanical issues with the pump itself. In such cases, the process department needs to safely hand over the pump to the mechanical department. To do that, the fluid contents inside the pump need to be drained and vented safely. Each pump is equipped with a set of drain and vent valves for draining and venting. The type of drain and vent and discharge location of drain/vents depends on whether the fluid service is flammable, hazardous, gas or vapor, etc. Such systems are shown in Figure 11.8.

For example, the non-hazardous liquid [Figure 11.8(a)] can be drained into an open drain system via a drain hub. All drain hubs from different pumps are collected into a common header and routed to a non-hazardous vessel for further processing.

The hazardous material, such as diesel or benzene, cannot be drained into the open drainage system. Such a hazardous chemical is collected in the closed drain header and processed in the hazardous drain vessel [Figure 11.8(b)]. Note that the material going into the hazardous drain system needs to be of low vapor pressure; otherwise, it can cause capacity and operational issues such as pipe hammering.

Hazardous materials with high vapor pressure cannot be sent to the closed drain system. For example, propane or butane materials have very high vapor pressure and exist in the vapor form at atmospheric temperature and pressure [Figure 11.8(c)]. All such services are routed to the flare system, where any carried liquid or condensed liquid can be removed through the flare drum, and the vapors can be safely burned in the flare.

Since the drain or vent system is not used regularly and used only when the pump undergoes maintenance, for safety, it is important to isolate the low-pressure drain and vent system from the potential high-pressure pump system. To achieve safety, a figure eight or spectacle blind in close position is added in the vents and drain systems.

As the pump could be rated for high pressure and the drain and vent system designed for a low-pressure rating, adding a piping specification break is necessary.

As the pumped liquid could be operating at high temperatures and the drain system typically operates at atmospheric temperatures, adding an insulation break on

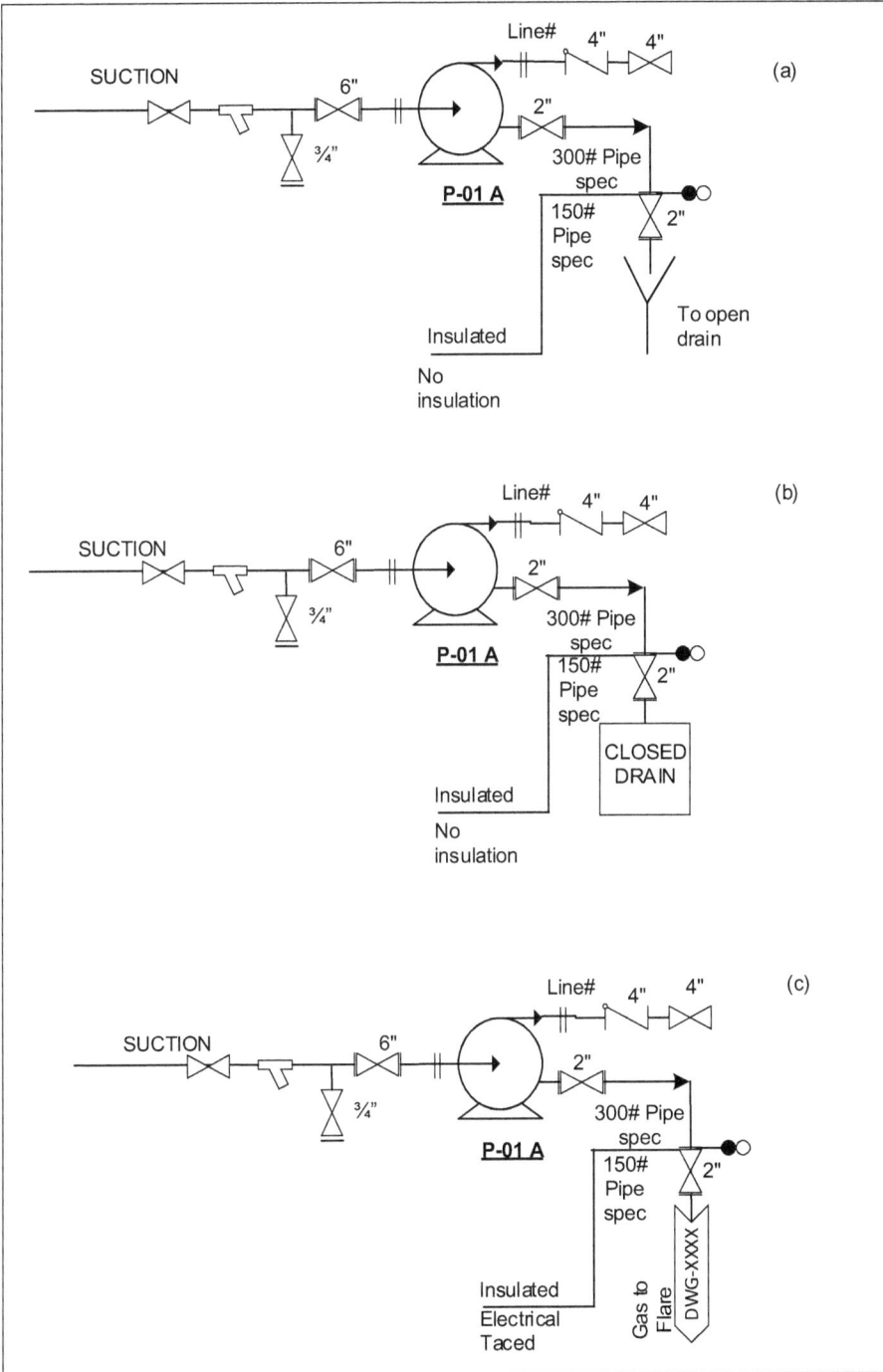

Figure 11.8: Pump drain and vent arrangement.

the drain or vent line is necessary to avoid any heat losses. For high-pressure liquids venting to the flare system, the vent lines could see freezing temperatures due to the Jules-Thompson effect. Such pipings are auto-traced with electrical coils to avoid freezing temperatures, which can lead to piping embrittlement.

11.9 Positive displacement pump

Positive displacement pumps have low flow and very high discharge pressure. They are used in process services such as chemical injection or dosing inside the process system. Pumps are small and used for chemical transport in very small quantities for high-head applications. Such a system is shown in Figure 11.9.

Every positive displacement pump has a pulsation dampener, both on the suction and discharge. The primary function of the pulsation dampener is to stabilize the vibration in the piping and to provide a continuous stable flow to the pump.

Since the positive displacement pump design can lead to very high pressures, these pumps are always installed with an internal relief valve. The operational hazard team from the plant and the engineering contractor may recommend adding an external relief valve in addition to the internal one, as a passive safety for the system. The discharge of the safety valve is routed back to the pump suction.

Since the volume collected from these pumps is relatively smaller than the centrifugal pump, the drips and drains during maintenance are collected in the pan and later sent to the hazardous drain system for processing.

11.10 Reactor system

The reactor system is a pressure vessel with a catalyst or layers of catalysts that alter the chemistry of the process fluid entering the reactor. The catalyst system is a very sensitive part of the processing unit as the catalyst can be deactivated by temperature fluctuations or unwanted chemicals present in the feed. The reactor system is monitored for temperatures and pressures for all such reasons. Such a system is shown in Figure 11.10.

The reactor pressure drop is measured by measuring the inlet and outlet pressures across the reactor bed and the difference in pressures is calculated and recorded. The pressure drop across the reactor bed is designed for a certain value; over time, it increases, as it fouls with fouling materials. If the pressure drop exceeds the allowed value, it impacts the catalyst's performance, and could create an additional burden on the feed pump, which may not be designed to handle the excess pressure drop.

The temperature inside the reactor fluctuates, depending on the fouling or coking mechanism. A localized increase in temperature can damage a portion of the catalyst,

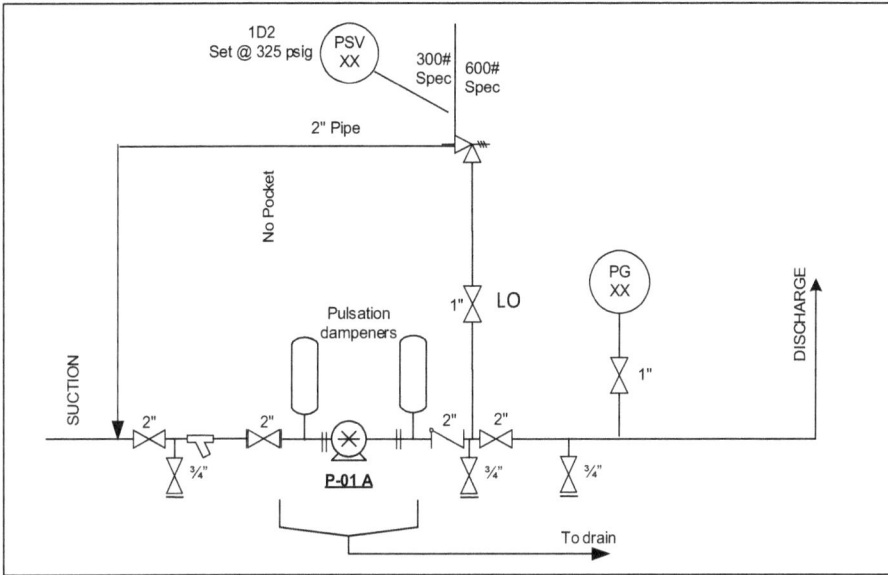

Figure 11.9: Details of a positive displacement pump.

impacting either the production or the specification of the product specified by the customer. In order to avoid such an impact, the temperature across the reactor bed is monitored using multiple thermocouples. Sometimes, the catalyst can have a runaway reaction, resulting in very high temperatures and damaging the vessel wall of the re-actor, causing hazardous material to leak into the atmosphere. For such scenarios, the highest temperature measured in the bed is compared in the comparator control block and can be defined as a trigger to shutdown the reactor system, based on the safety interlock system, as shown in Figure 11.10.

Some critical process equipment, such as the reactor, are equipped with pressure and temperature gauges on the inlet and outlet of the reactor to monitor the perfor-mance of the reactor visually when the operator goes for routine field checking.

During the start of the run, the catalyst is fresh, and it can perform efficiently, meeting the production and customer specifications. As the catalyst ages or comes to the end of the run, the catalyst experiences fouling. It may not function well, com-pared to its original design due to significant deactivation over time. For such reasons, the catalyst must be removed from the bed and replaced with a fresh catalyst every three to four years. For these reasons, a catalyst withdrawal nozzle is provided where the catalyst can be removed safely.

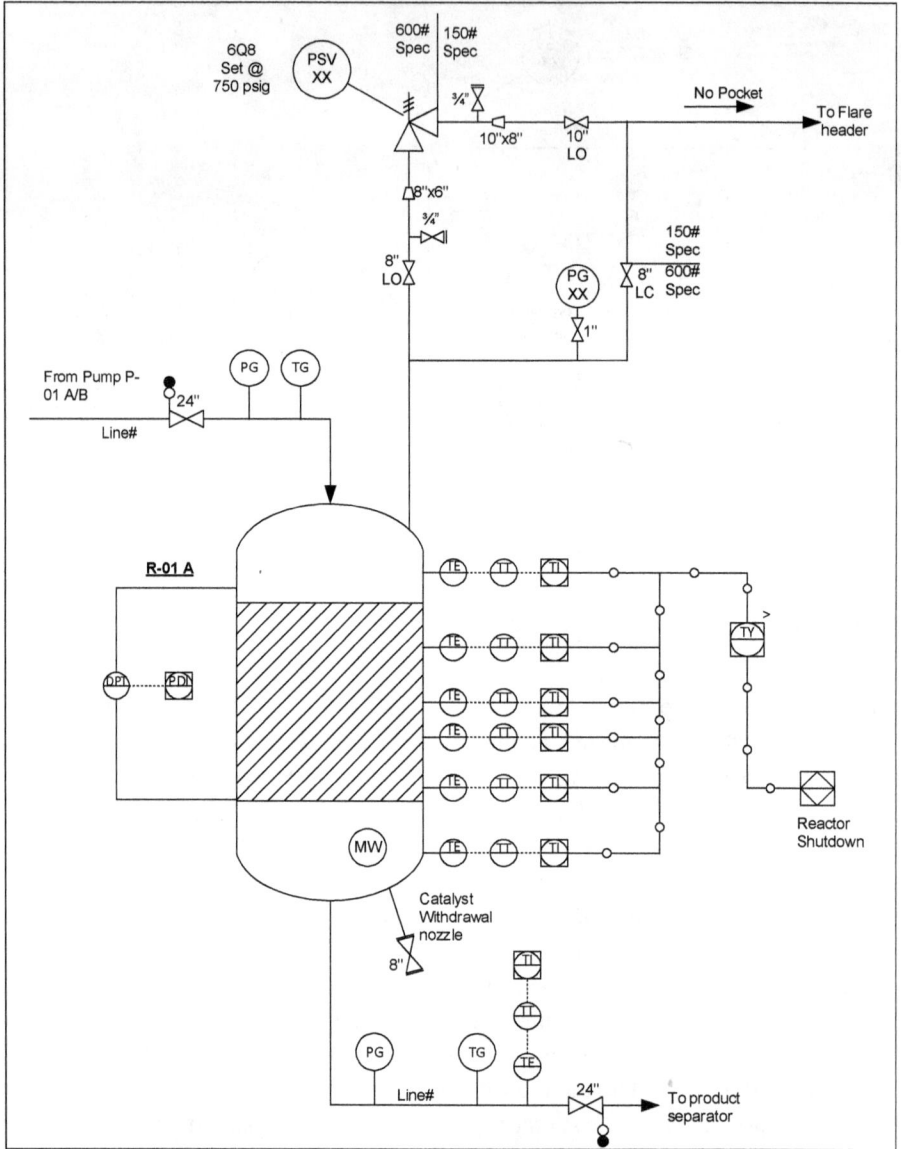

Figure 11.10: Reactor system.

11.11 Three-phase separator

Three-phase separators are used for separating gas, water, and liquid hydrocarbons, mainly in oil and gas operations. The separation of gas from liquids is required so that the liquid does not entrain into the gas system, which will be further compressed.

The separated water cannot have oil in it as it is a contaminant, and can be hazardous to the environment. The separated oil cannot have water due to the design of the downstream separation process equipment, and to avoid the formation of CO_2-water-based fouling in the pipelines. Such a system is shown in Figure 11.11.

The three-phase separator has several internals, each serving a particular purpose. These internals are a calming inlet baffle, mist eliminator, vortex breaker, and weir plate. The inlet calming baffle reduces the speed of liquid present in the gas stream and helps collect liquids from the inlet stream. As the gas molecules separate in the separator, it has a tendency to carry some fine mist of the liquid along with it. To avoid liquid carryover with the gas, a mist eliminator is installed on the gas outlet nozzle. The mist eliminator's design, shape, and pressure drop profile depend on the service, and are based on a desired pressure drop. The liquid collected in the vessel has two phases: water and oil. Oil being lighter than water, floats on top of the water and gets collected on the oil-side compartment of the separator using a weir plate. The separated liquid is connected to a set of pumps. In order to avoid creating a vortex inside the vessel, a vortex breaker is added that will ensure stable operation of a pump.

Maintaining a constant pressure inside the separator is important to provide a stable and efficient separation between the gas and the liquids. A pressure control valve is often added to the gas line, which maintains a constant pressure on the separator.

Both the water and oil levels in the separator need to be at a constant level to ensure a high degree of separation between oil and water. Also, this ensures that the gas does not escape into the liquid process, which can damage the equipment designed for liquids. In order to achieve the constant level in the separator, level controllers and control valves are added to the oil-side as well as to the water-side.

Some of the common parameters monitored in the separators are pressure, temperature, and level. Monitoring these parameters ensures that the efficiency of the separator is at its highest potential. For example, if the temperature in the separator is higher than anticipated, oil can escape into the vapor stream, and the separator can lose its efficiency quickly. So, by monitoring the temperature, the operator can anticipate the potential upset from the upstream process. A level gauge is typically installed on the oil- as well as water-side, in addition to the level transmitters. The level gauge installed helps assess the health of the equipment when the operator takes a routine field walk.

A pressure safety valve is added to the vessel to protect from overpressure scenarios such as liquid overfill, external fire, thermal expansion, blocked outlet, overpressure from the upstream process, power failure, and instrument air failure.

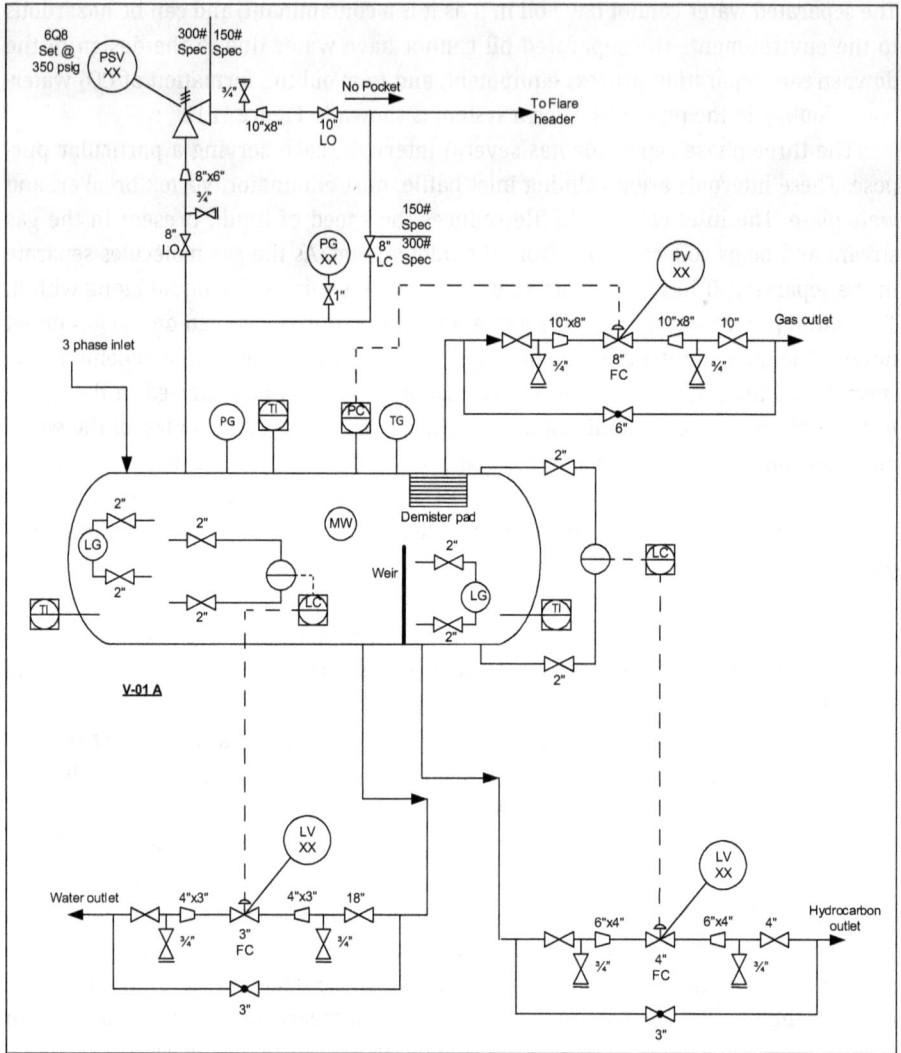

Figure 11.11: Three-phase separator.

11.12 Distillation column overhead receiver system

The overhead of the distillation column consists of gases or vapors that need to be condensed and collected in the overhead vessel before the liquid can be sent back to the tower as reflux and a portion of the liquid to the storage. The overhead system consists of a separation vessel and pumps. Such a system is shown in Figure 11.12.

Two sets of pumps are installed that take hydrocarbons separated from the overhead vessel and pump out reflux to the column, and the excess liquid goes out to the storage. The control valve located on the reflux line is designed as failed open, and the control valve sending liquid to the storage is designed as failed closed. The failed open position of the reflux control valve ensures that the cold liquid cools the column during an instrument air failure. Moreover, the control of the reflux control valve depends on the column's tray temperature. Essentially, the reflux control valve maintains the tray temperature by regulating the cold overhead liquid, which is being sent

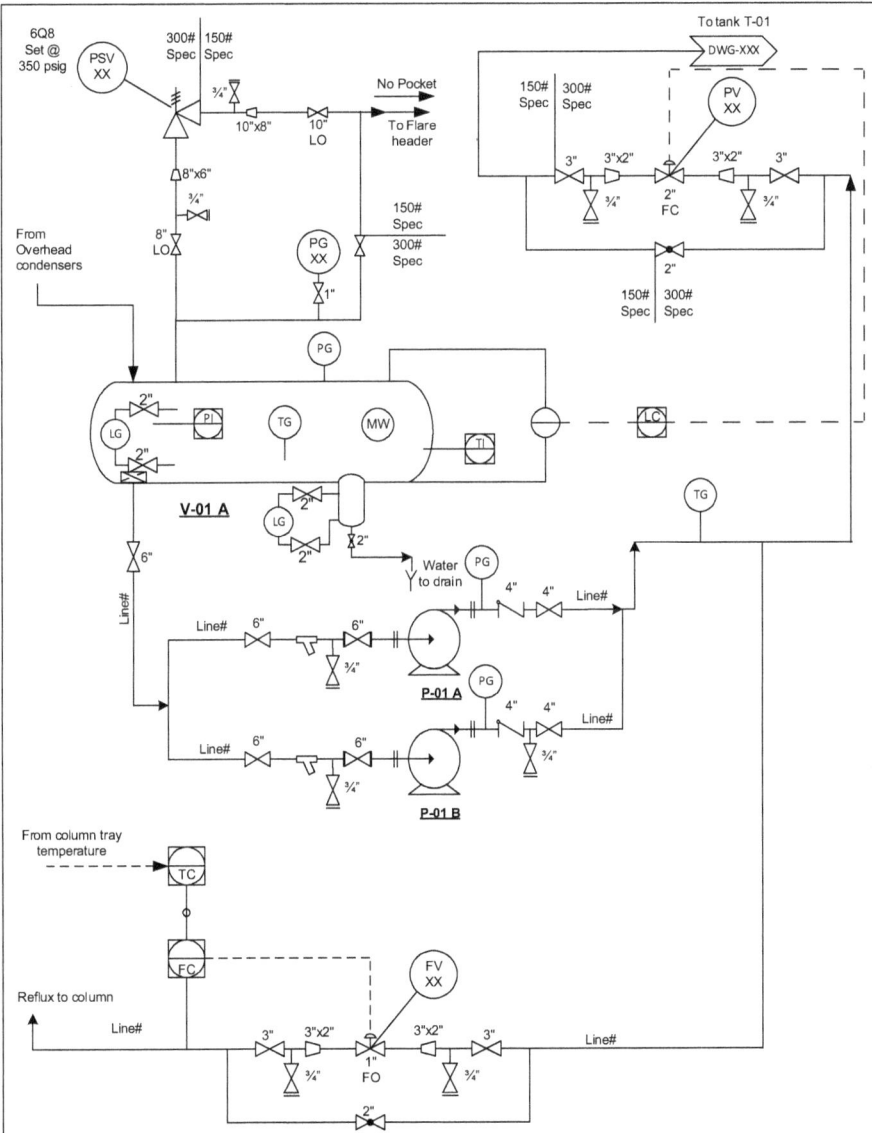

Figure 11.12: Distillation column's overhead receiver system.

to the column. Similarly, the failed close position of the control valve on the storage line ensures that the liquid inside the overhead vessel is not emptied completely, protecting the pump. Suppose vapors are collected, which could be uncondensed vapor at the operating pressure and temperature of the overhead vessel; in that case, these vapors need to be removed from the vessel, either to the flare or to the downstream collection vessel to avoid buildup of non-condensables in the systems.

As most of the overhead system has either the wash water system or extracts some water from the process streams, this water is collected in the boot section of the overhead vessel. Depending on the rate of water collection, the design may need to include a set of pumps or a simple line with a globe valve for water removal. As this water has some dissolved hydrocarbons and could have some H_2S, this water is further sent for processing to remove contaminants. Depending on the dissolved components in the water, H_2S or NH_3, the sour water needs treatment in the sour water-stripper units.

The relief scenarios for this vessel include external fire, block outlet, and failure of upstream condensers. The vessel is located 25 ft above to avoid a fire scenario. In addition to the relief scenarios, the vessel needs to be located below the overhead condensers to ensure free draining of liquids from the condensers to the overhead vessel.

11.13 Compressor knockout pot

There is always a liquid separation vessel before a gas compressor. The liquid separation vessel is also known as the compressor knockout pot. The vessel's primary function is to remove any trace liquids before the gas goes to the compressors. As the compressor cannot handle liquids and any trace amount of liquid can potentially damage the compressor, it is critical for the compressor knockout pot to work as efficiently as possible. Such a system is shown in Figure 11.13.

Process parameters such as pressure, temperature, and flow rate are measured around the compressor knockout pot. When the compressor does not receive an adequate amount of gas volume, the compressor can go into the surge scenario. In order to protect the compressor from this surge, the measured process parameters are sent to the surge controller, and appropriate action is later taken to avoid the compressor going into surge mode.

The inlet pipe is designed with internals that reduce the speed of liquid, and allow only the gas to escape. The demister pad installed in the vessel ensures almost complete removal of trace amounts of liquids. The inlet distributor pipe also aids in removing any entrained liquid from the upstream process.

The liquid acts as a seal between the gas and the liquid process systems. Since the collected liquid is in trace amount, a simple level control valve can control the level of the vessel. A high level in the vessel is not permitted as the high level can allow liquid to escape into the compressor, damaging it. Extremely low levels of a vessel are also

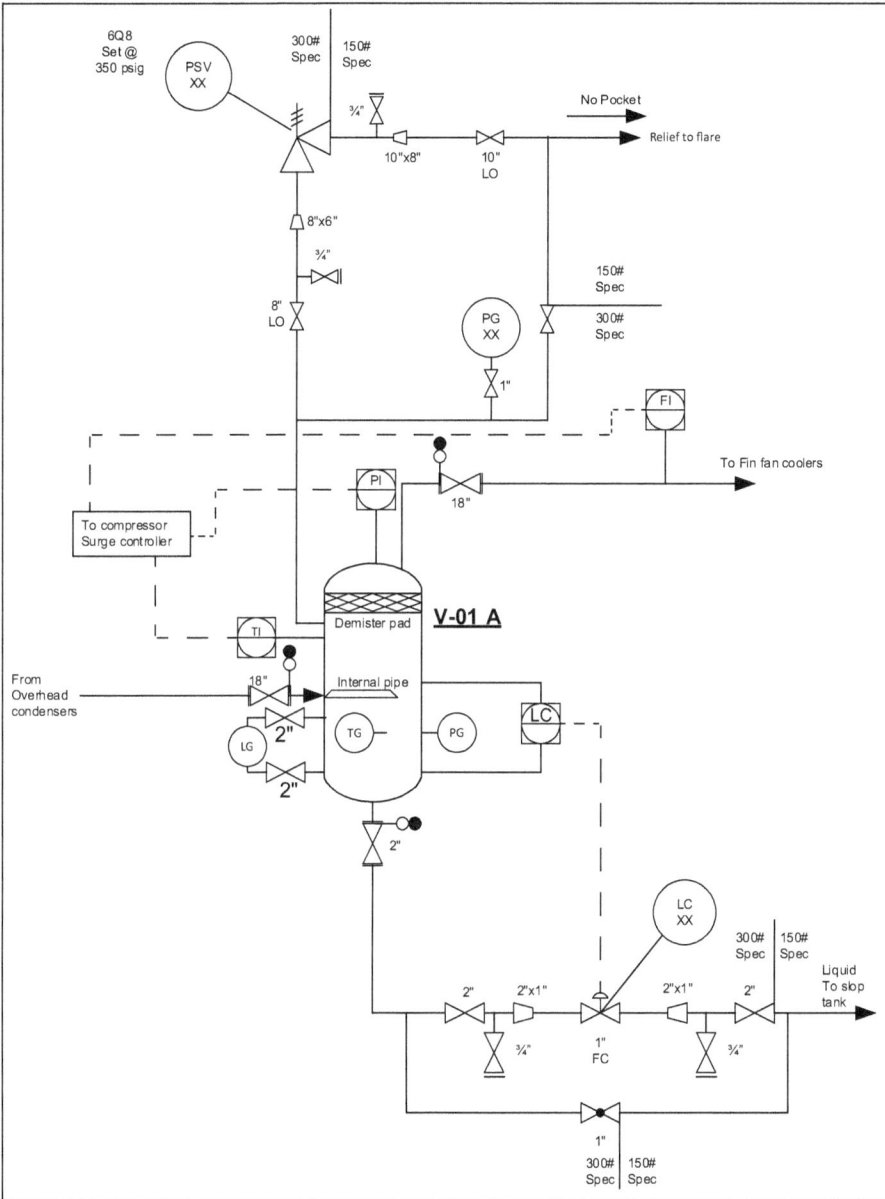

Figure 11.13: Compressor knockout pot.

not permitted as the high-pressure gas can escape into the low-pressure liquid system through the liquid line, and can overpressure the liquid system.

A relief valve is needed for this pressure vessel to protect from overpressure scenarios, such as the closure of gas or liquid outlet lines, fire scenario, overpressure from the upstream process, failure of the compressor, and failure of compressor surge control. The inlet takeoff for the relief valve is always below the demister pad. This is done to ensure a free path for gas or liquid straight to the relief valve. If the relief file is located above the demister pad, the demister pad can act as a restriction during a safety relief event.

11.14 Hazardous open drain drum system

Hazardous drains from different process systems, such as drains from the pump and the vessels or columns, cannot be sent to the atmosphere as these chemicals are hazardous. These hazardous chemicals have oil, hydrocarbons, metals, etc., and can harm the surrounding environment. Hence, it is necessary to collect these drains and process them. Such a system is shown in Figure 11.14.

A hazardous drain from the different systems is collected into a common header and sent to a drum, which is open to low-pressure (LP) flare through a pipe. It is necessary to have the headers free drain into the drum so that there is no pocket of liquid in the header piping. The pocket can create a restriction to the drain system. This vessel is called a hazardous open drain drum. The vessel is operated at near-atmospheric pressure and is located at the lowest level grade possible for gravity draining. The vent is sent to a safe location or LP flare. The weir present in the vessel separates out oil from water. It is important to provide adequate residence time for the oil molecules to separate from water. The customer or standard design guidelines specify the adequate residence time. The separated oil can be sent out to the slop system, where further oil can be recovered. The separated water can be sent out for further processing to remove any trace of entrained oil. Every pressure vessel has a manway where a person can enter the vessel and carry out any maintenance activities, if required, during a shutdown.

11.15 Shell and tube heat exchanger

Every chemical plant or a refinery has streams that require either cooling or heating. Both the cooling and heating steps require an area for an efficient heat transfer process. Shell-and-tube heat exchangers are most commonly used in the industry to exchange heat between two process fluids. In the shell-and-tube exchanger, one of the services is located on the tube side and the other one on the shell side in a countercurrent mechanism to maximize the heat transfer.

Process fluid allocation on the shell side provides a low-pressure drop. Process fluid allocation on the tube side allows the operation of a fluid that is fouling in nature.

In the example Figure 11.15, hot oil, a heating medium, is passed through the tube side, and a process fluid such as kerosene is passed through the shell side. Since the

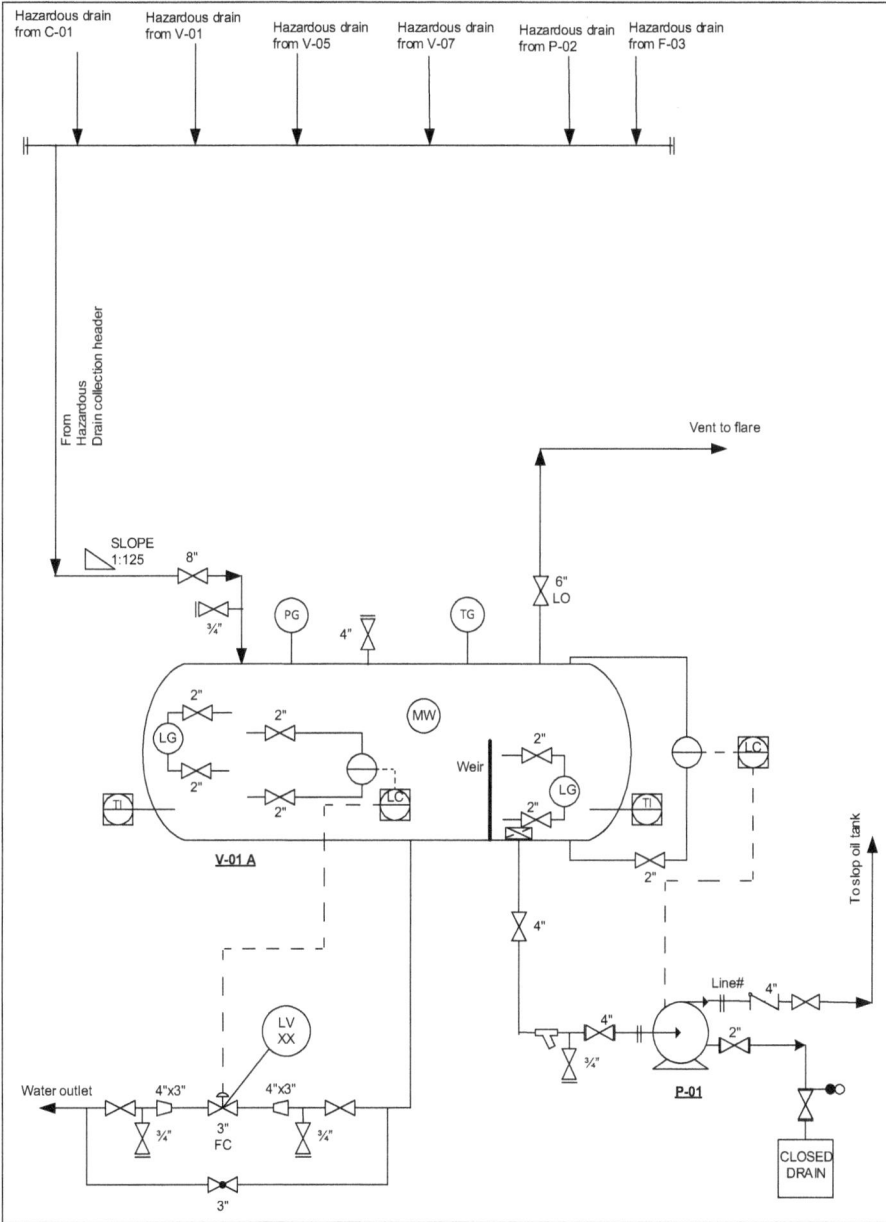

Figure 11.14: Hazardous open drain drum system.

kerosene fluid is cooler, it will pick up heat from the hot oil. The target temperature of the kerosene outlet of the exchanger is achieved by manipulating the flow rate of the hot oil in a flow-temperature cascade control. The inlet and outlet temperatures of hot oil and kerosene side are remotely monitored using temperature transmitters. Since the hot oil has a fouling tendency, the fouling rate can be measured by measuring the pressure drop across the tube-side. In order to do that, local pressure gauges are installed on the hot oil supply and return pipes. These pressure gauges are also important in monitoring the health of the exchanger. Similar to pressure, temperature gauges are installed on both the hot oil and kerosene sides at the inlet and outlet sides for monitoring the health of the exchanger during field walks. Both sides of the exchangers are always provided with isolation and drain valves. The isolation valves can be closed, and the exchanger can be given to maintenance for any clean-up or other maintenance activities. For maintenance activities, the drain valves, which are ¾" in size, are used to empty

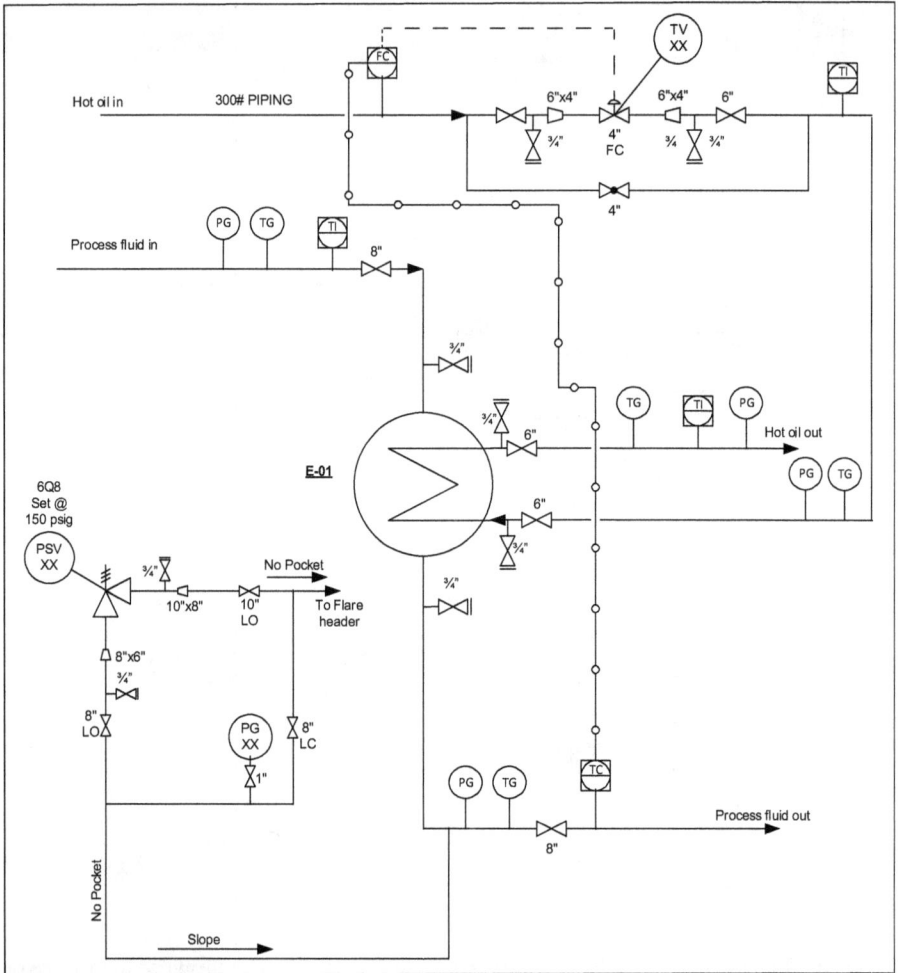

Figure 11.15: Shell-and-tube heat exchanger arrangement.

out the contents of the heat exchanger. Sometimes, hazardous vapors are handled in the exchangers, which cannot be vented to the atmosphere. For such systems, a dedicated line, routed to the flare, is needed to ensure that the vapor contents from the heat exchangers can be safely vented, and the exchanges can later be given for maintenance.

While designing a heat exchanger, if the low-pressure side design pressure is not designed for at least 10/13th value of the high-pressure side design pressure, the low-pressure side needs a relief valve to protect against tube rupture arising from the high-pressure side. In the example Figure 11.15, the kerosene-side has a relief valve set at 150 PSIG. In the example explained, in addition to other scenarios, tube rupture is one of the sizing scenarios.

Rupture pin valve design for rapid depressurization (e.g., CW vs. hydrocracker HP reactor fluid at 2,000 psig) is not economical to design to the 10/13th rule: Sometimes, it is not possible to design for the 10/13th rule on the low-pressure side as the economics do not make sense. For such situations, relief valves are always needed. Another example is cooling water on the tube-side and the hydrocracker high-pressure reactor fluid operating at 2,000 PSIG. In this example, the cooling water-side cannot be designed for 2,000 PSIG as it will not be economical. A relief valve on the cooling water-side might be a better design in such cases. For such high-pressure differential systems, a rapid relief response is often needed, and a rupture pin valve design is often needed.

11.16 Fin fan heat exchanger

Fin fan exchangers use air-based cooling. A forced draft fan driven by a motor pushes cool air through a fin fan bundle. The fin fan bundle is nothing but many tubes arranged with thin fins that provide high heat transfer. The fin fan can provide a 25% natural convection duty when the motor is not working. Fin fan exchangers are commonly used in the overhead condenser service for condensing column vapors, generally to 140 °F temperature. They are primarily used on column overhead and remote locations where cooling water is unavailable.

These fin fan exchanges have a larger footprint than the shell-and-tube exchangers. Some fans need to be turned off during colder climates. The duty of the fin fan exchanger is controlled either by louvers that adjust the angle of airflow coming to the bundles or by speed control using a variable frequency drive motor (VFD). As shown in Figure 11.16, the outlet temperature of the fin fan is controlled by regulating the speed of the motors.

Symmetric piping and hydraulic symmetry between two bundles are sometimes necessary to distribute the vapor or liquid to ensure uniform coverage of the cooling area. Drain valves and isolation valves are also added to the exchanger system, which provide necessary isolation and draining, as required, during maintenance.

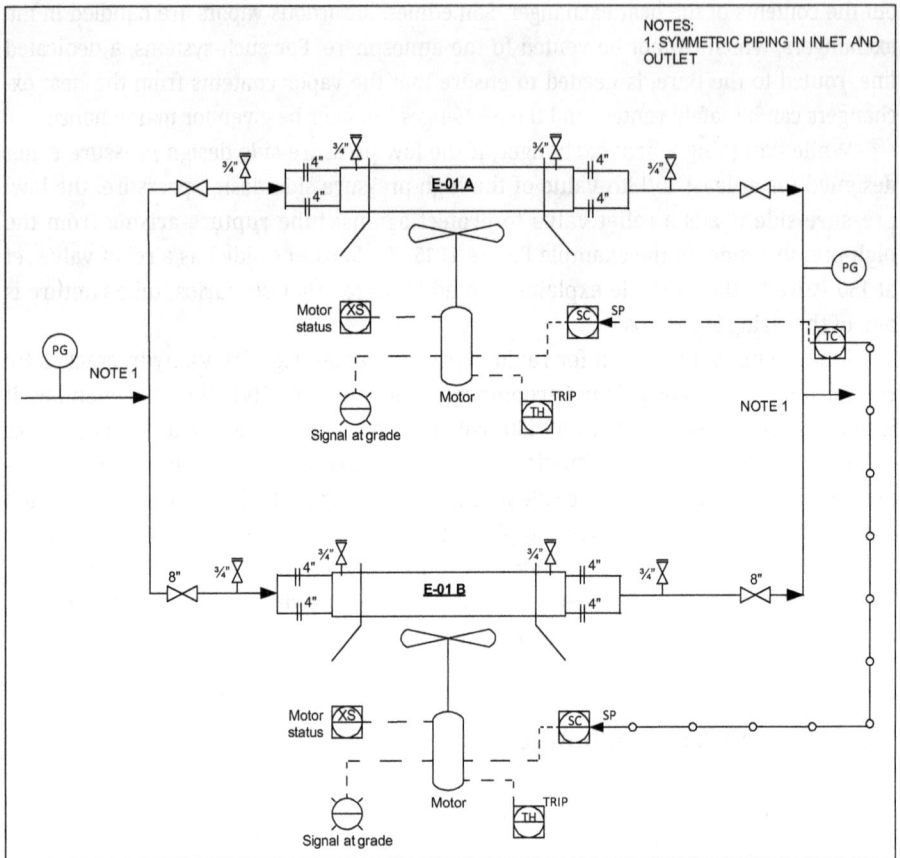

Figure 11.16: Fin fan heat exchanger arrangement.

11.17 Double-pipe heat exchanger

Similar to other exchangers, double-pipe exchangers are easier to design and consist of one pipe inside another. Hotter fluid is passed through one of the pipes, and the cooling fluid is passed through the other pipe. This exchanger is used for a very small-duty application. Also, double-pipe exchangers utilize smaller plot space.

Figure 11.17 shows a double-pipe heat exchanger. The product from the column is cooled using cooling water on the other side of the exchanger. The control valve on the process-side outlet controls the level inside the column. Similar to other exchangers, a pressure safety valve may be necessary on the low-pressure side to protect against tube rupture scenarios.

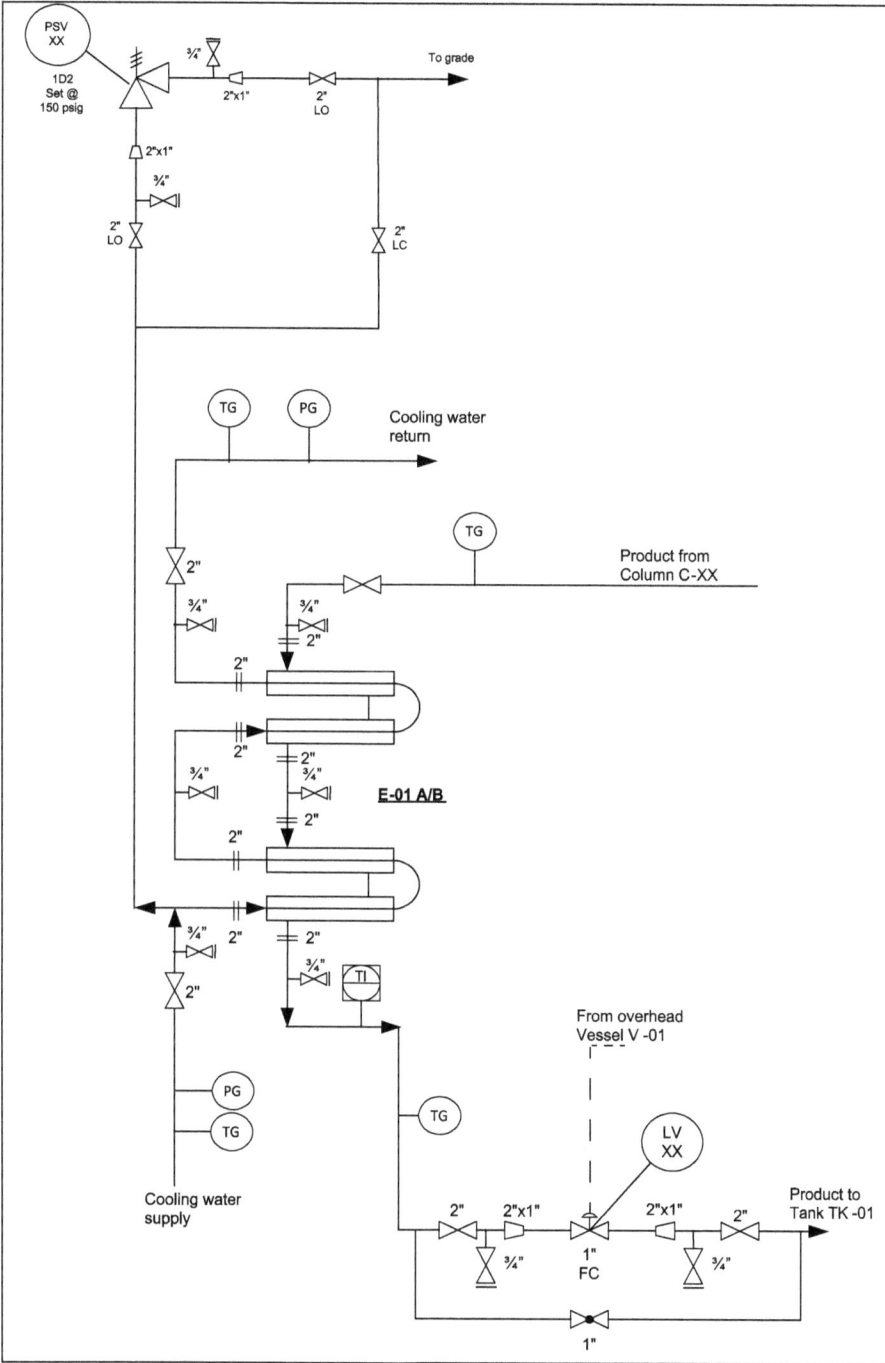

Figure 11.17: Double-pipe heat exchanger arrangement.

11.18 Kettle-type of heat exchanger

The kettle-type heat exchanger has a tube bundle where process fluid is cooled using cooler fluid on the shell side. The shell-side arrangement differs greatly from a typical shell-and-tube exchanger. The shell-side of the exchanger is used for generating steam using boiler feed water (BFW). The steam produced is routed to the steam header in the plant. BFW flow rate regulates the level on the shell-side of the kettle. Over time, solids are deposited at the bottom of the exchanger, and a blowdown of the shell-side may be necessary to eliminate the solid from the exchanger. A fixed blowdown is provided using a flow control mechanism. The shell-side of the exchanger is protected from over-pressure scenarios such as external fire, abnormal heat, blocked outlet, etc. Figure 11.18 shows an arrangement of kettle-type heat exchanger.

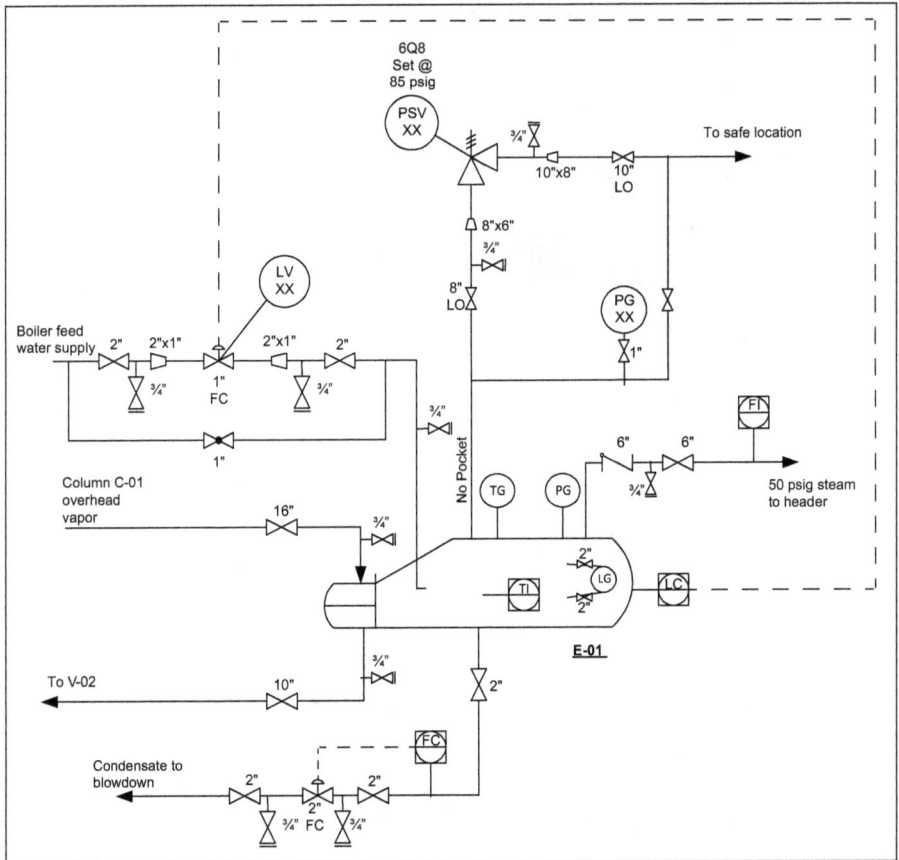

Figure 11.18: Kettle-type heat exchanger arrangement.

11.19 Vertical type of reboiler

The shell-side of the heat exchanger may be arranged in a vertical position for the vapors to escape easily into the process. A common application is a column reboiler. Liquid from the column enters the bottom of the shell-side of the shell-and-tube exchanger, and the vapor generated is returned back to the tower. The heating medium, usually on the tube-side, can be steam or hot oil. The column temperature of a particular tray is controlled by regulating the hot oil flow. It is necessary to keep the pressure drop on the process side as low as possible to achieve the highest possible circulation rate required for the vapor-liquid traffic inside the column. Forced pump circulation system are typical applications for larger heat duties. Similar to other exchanges, the hot and colder sides have isolation and the block valves necessary for isolating the reboiler, and draining the contents during maintenance. Local pressure gauges and temperature gauges are also installed on both sides of the exchanger to monitor fouling and assess the exchanger's health during field walk. Figure 11.19 shows such an example.

11.20 Ejector system

The ejector system consists of shell-and-tube condensers, an ejector body, and other piping. The ejector is used to create vacuum in the process system. The ejector system uses 50 PSIG steam as a motive fluid. When the steam enters the ejector nozzle, a low vacuum is created, and the process gas from the processed vessel is removed. The hot gases are later condensed in the water-cooled condensers. Uncondensed vapors are sent for downstream processing. The condensed liquids are typically sent to the hot well for further separation of water from oil. A spare ejector system is always recommended as the ejector can malfunction. Depending on the level of vacuum needed in the process vessel, more than one ejector stage might be needed. Removing impurities or foreign objects in the steam system is important to avoid blocking the ejector nozzle. The steam system is always equipped with strainers or a knockout drum to achieve such removal. A steam trap is also required before the strainer. An example of a steam ejector system is shown in Figure 11.20.

11.21 Utility station

Every plant has several utility stations. The function of a utility station is to provide necessary utility air, low-pressure nitrogen, low-pressure steam, and utility water for cleaning, purging, and airing out the vessel or piping during startup or shutdown events.

The utility air is taken from the air header tied to the compressor discharge before the air dryer. As moisture-free air is not necessary for utility requirements, the quality of the utility air is sufficient for most of the startup or shutdown operations.

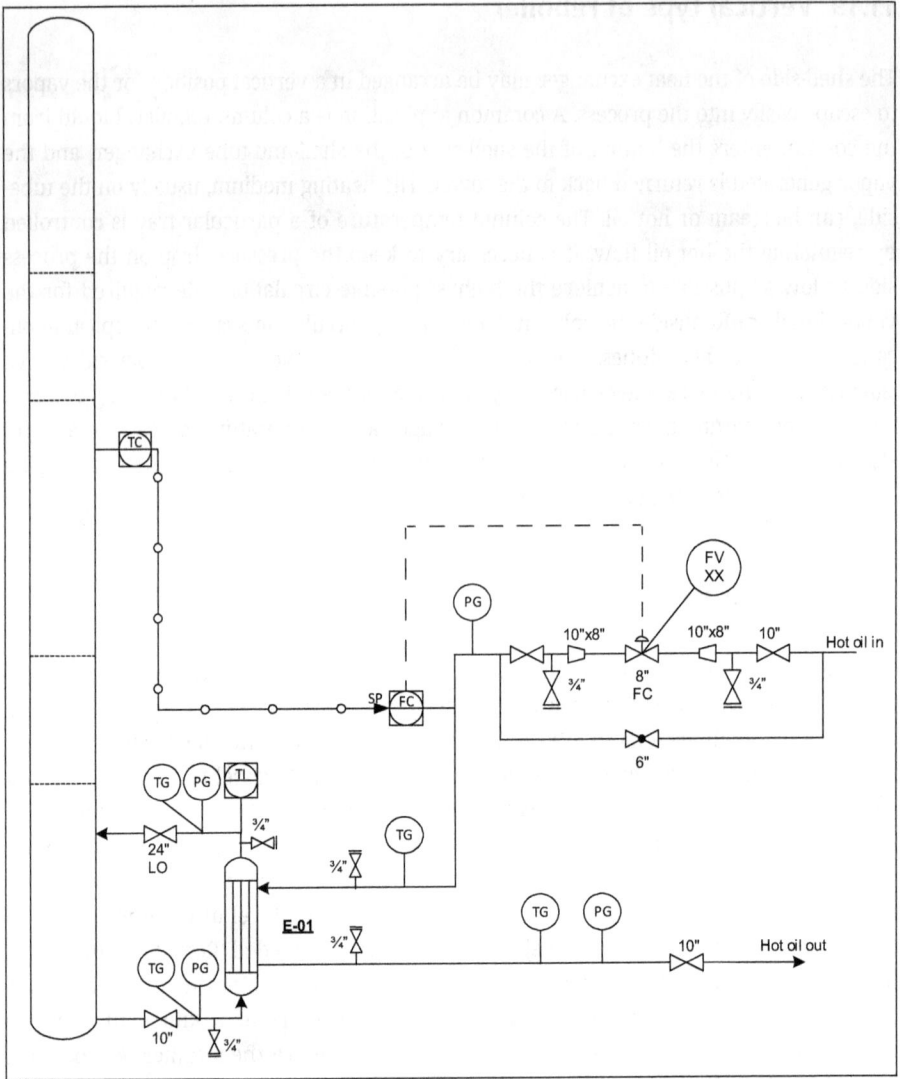

Figure 11.19: Vertical-type reboiler arrangement.

Utility air is commonly available between 80 and 90 PSIG. Low-pressure nitrogen is taken from a high-pressure nitrogen system. High-pressure nitrogen, commonly available at 300 PSIG, is let down to 50 PSIG, distributed to users, and connected to low-pressure nitrogen utility stations. The low-pressure steam, available at 50 PSIG, is produced within the plant by letting down the medium-pressure steam of 250 PSIG. Utility water is clarified water, and it may not be purified, but it is sufficient for shutdown and startup activities.

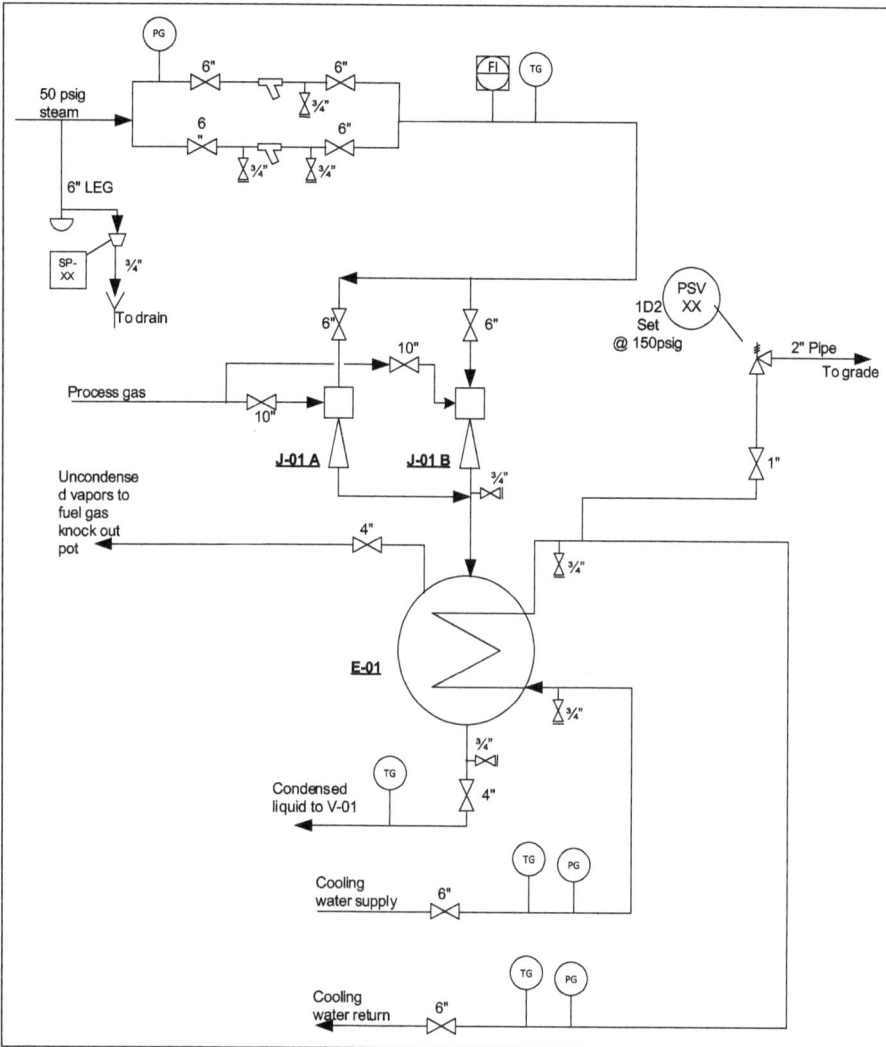

Figure 11.20: Ejector system arrangement.

All these utility connections are hard-piped and brought to a station. Each utility line has a set of block valves and a check valve of 1-inch size. The check valve ensures that there is no backflow into the utility headers, which could contaminate the utility system. Typically the 1" pipe is reduced down to the actual hose size, which is ¾" in size.

Sometimes, there may be a special requirement for hot utility water at 200–220 °F. These temperatures are obtained by mixing low-pressure steam with the utility water using globe valves and by monitoring the temperature. Both these streams are passed through a hot water mixer, ensuring adequate mixing of the utility water and the low-pressure steam. Figure 11.21 shows a typical arrangement of a utility station with a hot water mixer.

For certain plant areas, where equipment may not be available, only air and water might be needed as utilities. In such cases, nitrogen and low-pressure steam connections are not required.

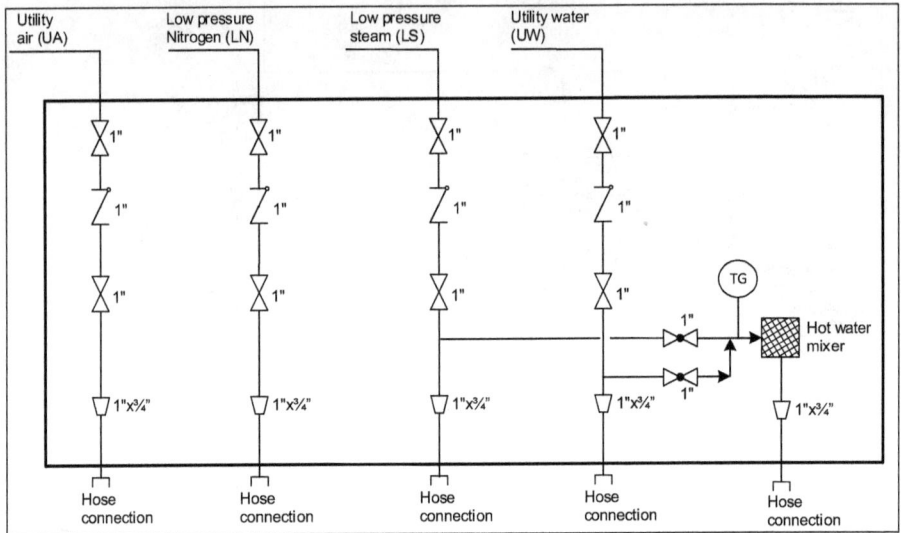

Figure 11.21: Typical utility station.

11.22 Storage tank with a pump system

A storage tank is not a pressure vessel but a tank rated for very low pressure. The primary purpose of the storage tank is to hold the necessary inventory for the desired amount of time. The pumps, which are connected to the storage tank, transfer the material to the loading dock or railcar as a shipping product during the demand. The type of storage tank and roof type depends on the fluid service handled inside the tank. The fluid material with a low vapor pressure, such as diesel, can be simply stored in a floating roof tank. In contrast, fluid services such as butane with somewhat higher vapor pressure must be stored under pressure and typically in spherical-shaped tanks, where the pressure inside the sphere is uniformly distributed.

The fill line for the tank comes straight from the process area where the process product is made. As the material in the tank is constantly added or removed, the change in level in the tank affects the pressure inside the tank. A pressure and vacuum relief valve (PVRV) is always installed on such tanks. Emergency doors or vents are installed on the tank to account for external fire scenarios. Two different types of level instruments are always installed on a tank. The most used level instruments are differential pressure (DP)-type and radar-type instruments. Temperature is also an-other important parameter to monitor for a tank. Both temperature gauges and tem-perature transmitters are installed to monitor the temperature of the liquid. In order to protect the pump from going below the pump's minimum flow requirement, a min-imum circulation line is always added and routed back to the tank. It is important to add a dip pipe internal to the minimum circulation line and the inlet fill line to avoid any flashing or to avoid disturbance to the liquid level. A set of pumps transfers the liquid to the user. A control valve located downstream of the pump controls the level of the tank by modulating the flow rate through the pump.

Water boot may be necessary to remove the settled water from hydrocarbons. Adding water boots to the storage tank is common in refineries. Where possible, an overflow line or overflow protection is added to the tank to handle overflow scenar-ios. The overflow protection can only be added for non-hazardous services such as water. The overflow protection cannot be added for hazardous services such as hy-drocarbons. An example of a water storage tank with a set of pumps is shown in Figure 11.22.

11.23 Storage tank blanketing system

Storage tanks that store hazardous chemicals or hydrocarbons cannot be released into the atmosphere. When the liquid is filled into the tank, it displaces the vapor space, creating pressure inside the tank. When the liquid is emptied out of the tank, vacuum is created inside the tank. A split-range control system is most commonly used to handle these scenarios. It consists of two control valves operating based on a single pressure controller, which splits the control action into 50:50. The low-pressure nitrogen makes up the necessary pressure during the emptying-out process. The line to the flare and the control valve removes any excess pressure from the vessel during the fill event. It is important to note that a check valve may be needed on the nitrogen line to avoid any backflow of chemicals or hydrocarbons into the nitrogen utility sys-tem. The line going to the flare and the nitrogen line are always designed so that any accumulated vapor or droplets can freely drain back into the tank. It is important to note that the flare control valve is always designed for a failed-open position to avoid any excess pressure buildup inside the tank. Similarly, the nitrogen control valve is designed for a failed-close position to avoid adding excess pressure to the tank. Such a system is shown in Figure 11.23.

11.24 Storage tank mixer and heater

Oftentimes, the storage tank stores some heavy hydrocarbons, such as heavy vacuum gas oil. When this material is kept at atmospheric temperature, it tends to solidify. In

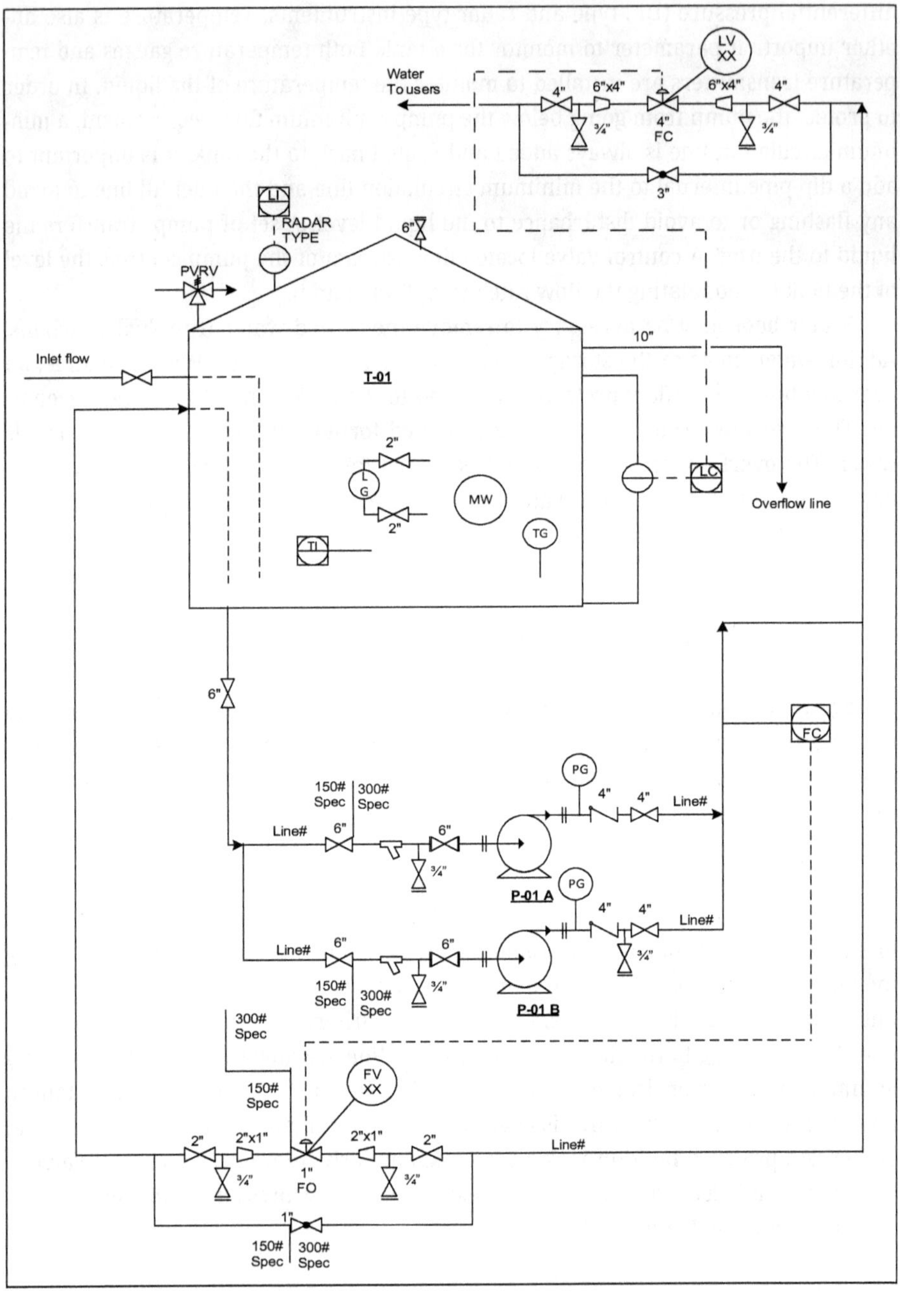

Figure 11.22: Storage tank with a pump system.

order to keep the material in a liquid state, heating coils are often used. The heating coils are supplied with a low-pressure steam or low-pressure condensate as a heating

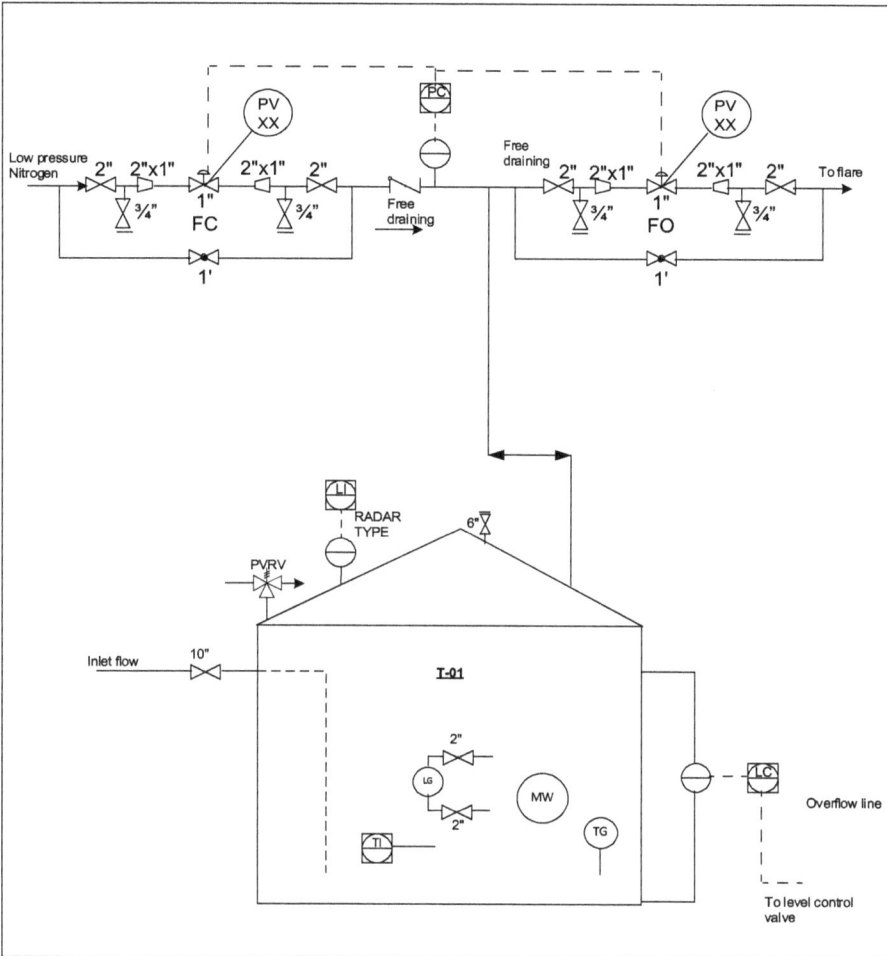

Figure 11.23: Storage tank with nitrogen blanketing system.

medium. The target temperature inside the tank is achieved by modulating the steam or the condensate flow rate.

Moreover, for the heating coils, it is necessary to keep the liquid inside the tank circulating to avoid any solidification during pump-out operations. The condensate collected after the heating is removed using a steam trap and can be drained into the drain hub or collected into a header and recycled back for reuse. Also, during pump-out, to meet customer specifications, thoroughly mixing the tank is necessary to ensure that a uniform product is provided to the customer. A set of eductors achieves the mixing. The pump provides the pressure differential for the eductor, and a line back to the tank provides the necessary circulation flow. It is important to account for

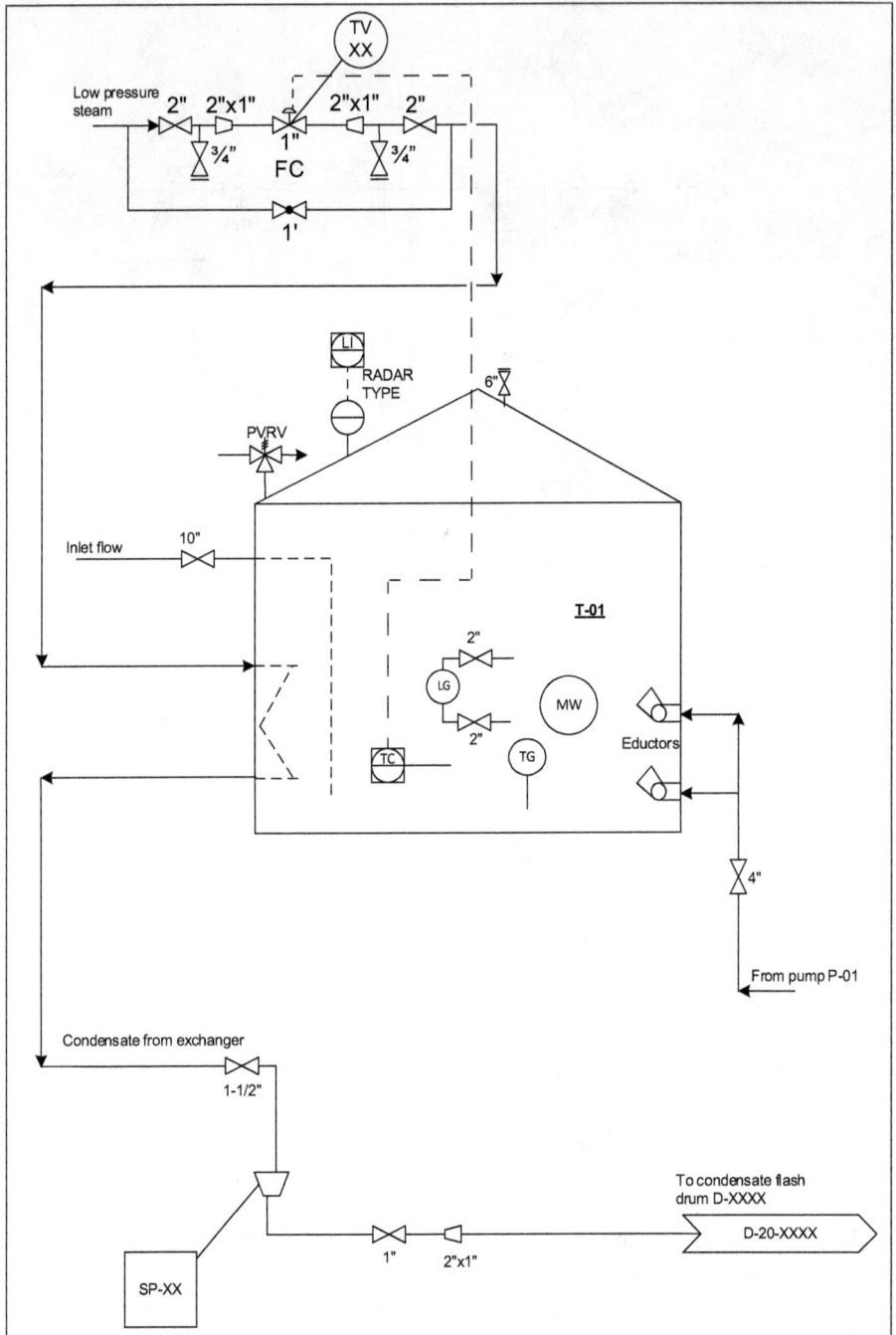

Figure 11.24: Storage tank with mixer and heater.

the pressure differential of the eductor during the pump design. An example of such a system is shown in Figure 11.24.

11.25 Distillation column

Every chemical plant has several distillation columns. Each column may be different in design and operation. Each customer may have a different set of guidelines for depicting a distillation column on a P&ID. A simple representation of the distillation column is shown in Figure 11.25. It consists of a pressure vessel with a smaller diameter and a significantly larger tangent-to-tangent length. The column could be raised above the grade level using a skirt height to create a net positive suction head for the pump. Typically, trays are numbered from top to bottom and are shown inside the pressure vessel. Different nozzles connecting to and from the column as also their sizes are shown. Some of the common nozzles are feed, pump minimum circulation line, vapor outlet line, reflux line, reboiler inlet line, column bottom line, and several instrumentation connections of 2". Pressure drop across the column is measured using a pressure differential transmitter to check the health of the column. If the pressure drop is higher than anticipated, there could be a malfunction inside the column, or fouling could be present. As the temperature inside the tower changes throughout the trays, lowest at the top and highest at the bottom of the tower, it is necessary to monitor the temperatures to assess the health of the column operation. To monitor and assess, several temperature transmitters are added at different sections of trays, as shown in Figure 11.25. Figure eight or spectacle blinds are often necessary to completely isolate the distillation column. Spectacle blinds help isolate the column positively from other processes during shutdown events. Local temperature and pressure gauges are installed on the column to assess the health of the column during field walk.

A distillation column has more than one tray inside it. A customer may prefer to add other column details, such as downcomers, tray types, and other special details, as needed.

11.26 Pig launcher

Offshore systems are typically installed with pig launchers and pig receivers. These equipment help in cleaning up the process piping using a pigging system. Motor operation valves (MOV) help in isolating the main process plant for the pig launching or receiving operation. A pig launcher consists of a pressure vessel-looking equipment. The pig is slightly smaller in diameter compared to the actual pipe size being pigged. The pig launcher works based on a certain pressure differential. The pressure differential can be achieved by sending high-pressure nitrogen or utilizing the well-head gas pressure. Once the pig enters the pipeline, it cleans up the inside of the pipe, and unwanted material from the piping can be received from the other end of the pig receiver.

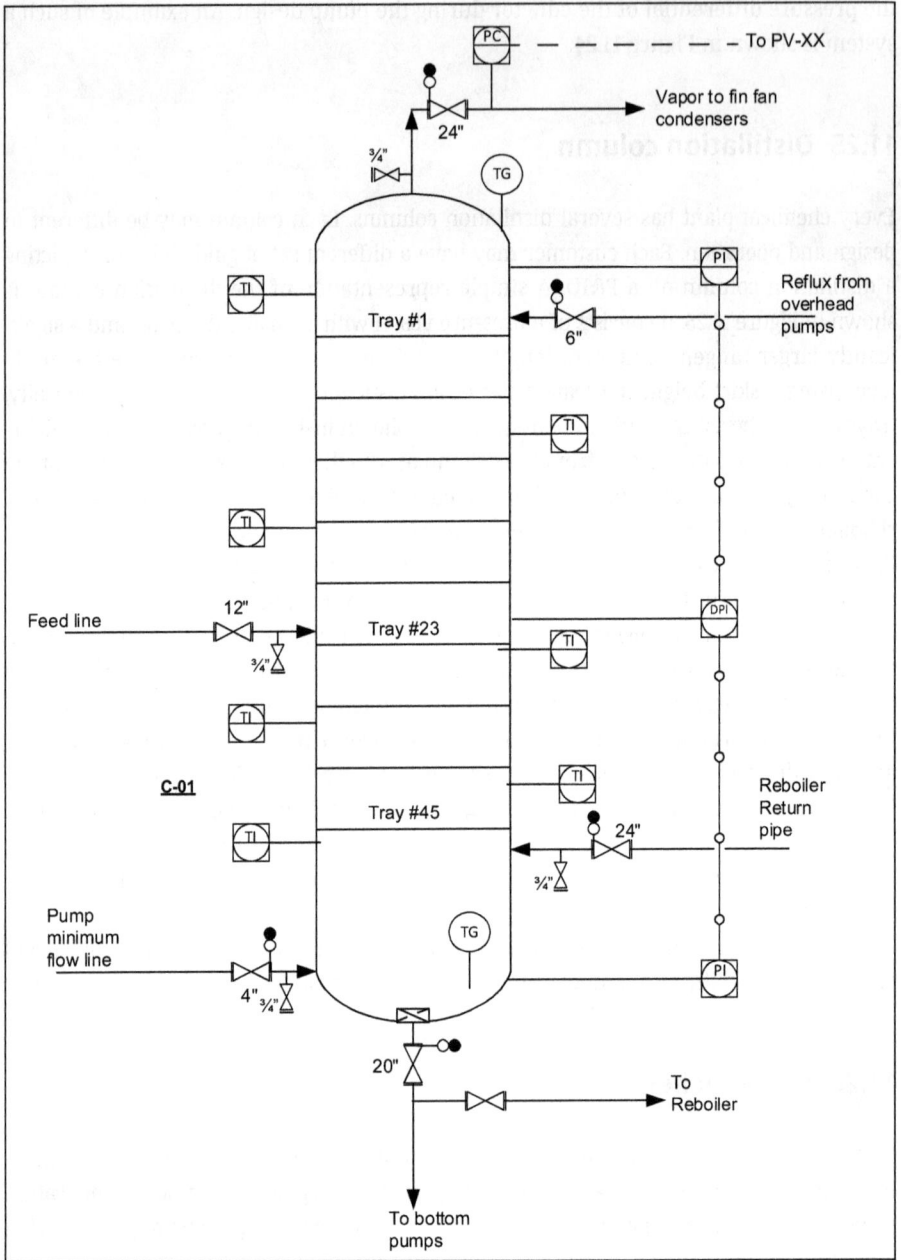

Figure 11.25: Distillation column.

Pig launchers are always equipped with a safety valve, equalization line, sand detection instrument, pressure and local temperature gauges, and drain lines and valves. The safety valve protects the pig launcher or receiver from overpressure scenarios. The equalization line ensures that once the pig is received, there is equal pressure on both sides of the pig. Local pressure and temperature gauges are used to monitor the process parameter on the pig launcher or receiver. Once the pig launching or receiving operation is completed, the contents from the pig launcher or receiver can be drained using drain valves and can later be collected in the slop tank. Any pig launcher or receiver vapor can be routed to the flare through the relief valve bypass line. The relief valve bypass line has a restriction orifice designed to restrict high-pressure vents going straight into the flare system, which could damage the flare system. Piping specification breaks are needed to isolate a high-pressure system from the low-pressure system, also shown in Figure 11.26.

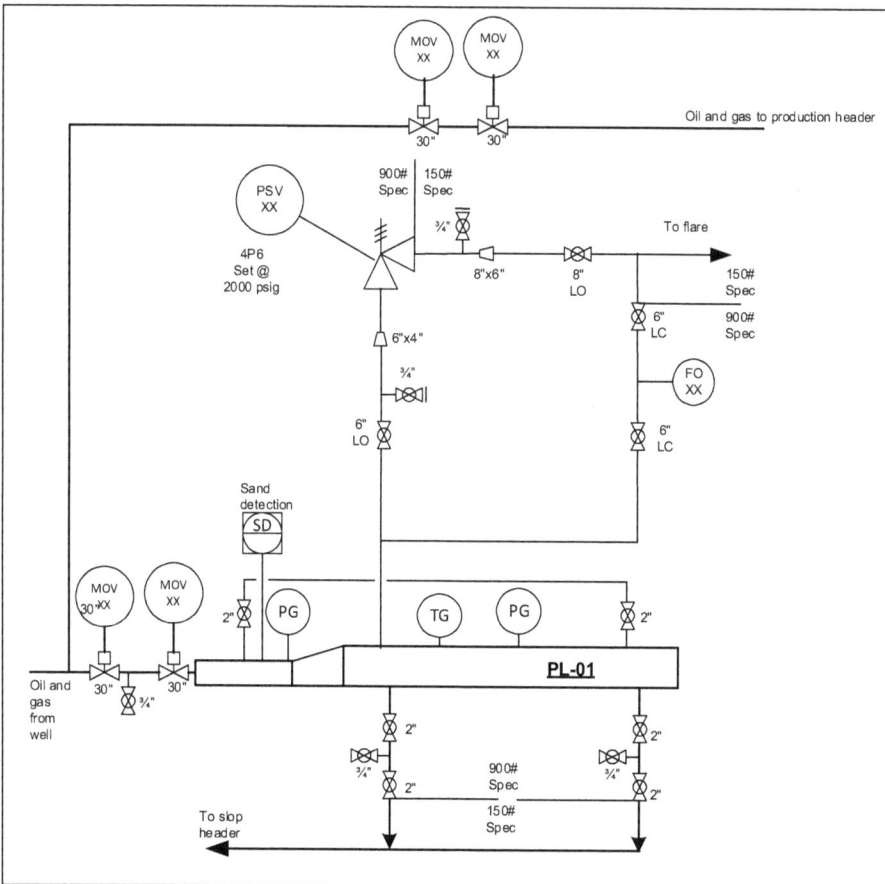

Figure 11.26: Pig launcher.

11.27 Slug catcher

Slug catchers are installed to collect any liquid slug from the offshore pipeline. In the off-shore processing plant, a slug or liquid can come toward the processing unit and can cause process upsets in the three-phase separator vessel. For such reasons, the gas with a slug of liquid is routed through the slug catcher, where the slug of liquid can be removed, and the gases can be routed to the process separator. Slug catchers are very similar to three-phase separator vessels, which are horizontal in orientation. The primary differ-ence between three-phase operator and a slug catcher is that the volume of the slug

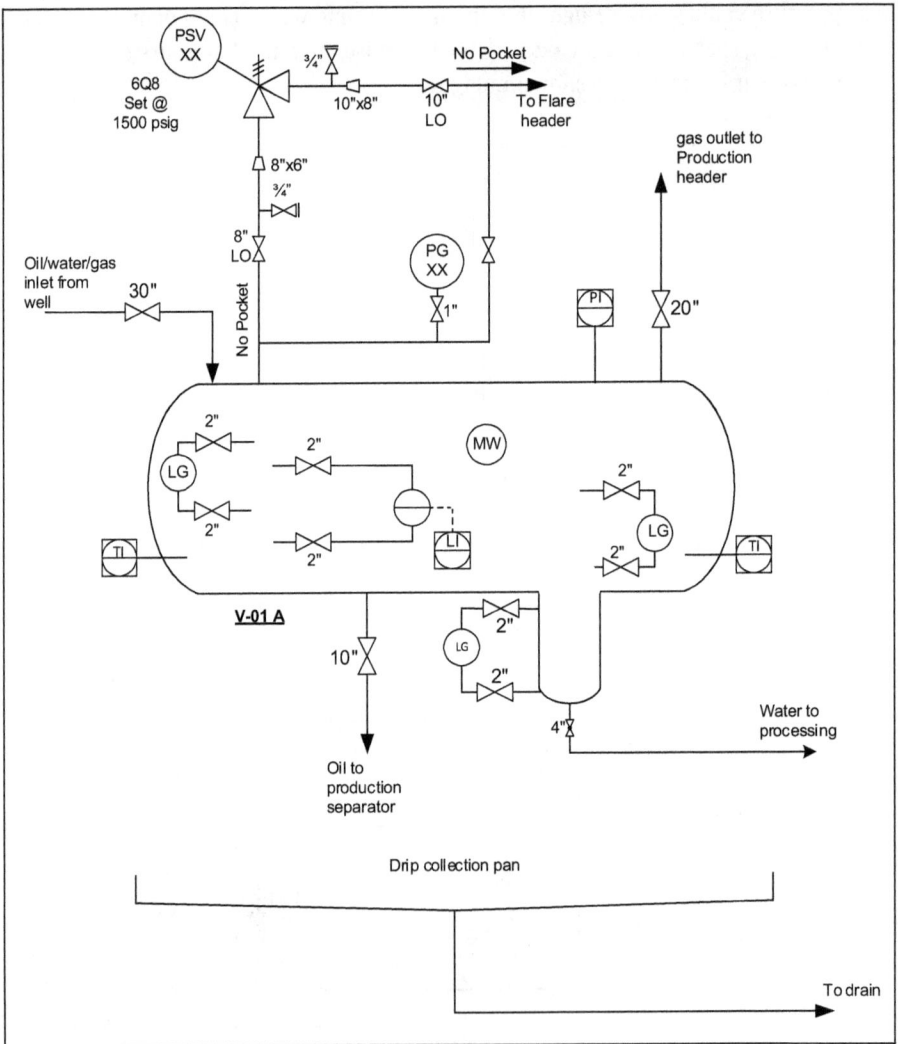

Figure 11.27: Slug catcher.

catcher is higher than the three-phase separator. Also, the slug catcher is also designed to remove any collected water through the boot system. During the maintenance of the slug catcher, liquids can be collected in the drip collection pan and drains, which can be drained to the hazardous drain vessel for further processing. Figure 11.27 shows an example of slug catcher.

11.28 Slop oil collection header

Slop oil is waste oil from the processing unit, which needs to be rejected from the unit. Slop oil can be a mixture of hydrocarbons and water. It can come from various sources, such as a collection of slop from draining a vessel or cleaning a pump. Different slop oils from different sources are collected in a common header or sub-headers and routed into the slop oil tank system. It is necessary to ensure that the pipe and the headers are designed to free-drain appropriately into the slop oil tank, ensuring no pocket inside the header system. Also, adding some slop oil connections for future expansions may be a good idea. The header system of slop oil is shown in Figure 11.28.

11.29 Liquid coalescer

Sometimes a liquid hydrocarbon stream can have trace amount of water, which needs to be removed before hydrocarbon liquids are pumped out to the customer. This is done by using a liquid coalescer. The coalescer involves coalescing filter elements that help collect the water molecules and separate the hydrocarbons from the water. As shown in Figure 11.29, two zones in the coalescer separate water from the oil. The collected water is sent for further processing, and the water level inside the liquid coalescer chamber is controlled by regulating the water flow. A pressure safety device protects the coalescer from any overpressure scenario such as external fire, blocked outlet, or overpressure from upstream or downstream systems. Temperatures and pressures are monitored in the coalescer to assess the health of the coalescing operation.

11.30 Sump system

When a plot space constrains the process plant, a sump system is usually recommended. The system is located underground and collects the chemicals or oils from different sources, such as pump drains. As the level in the sump tank increases, a set of submersible pumps transfers the liquids to the desired location. As the source may have solids and contaminants, it is necessary to add a set of filters on the pump discharge to remove any solids and to avoid damaging the downstream equipment and piping. The pressure differential across the filter is always measured to assess the

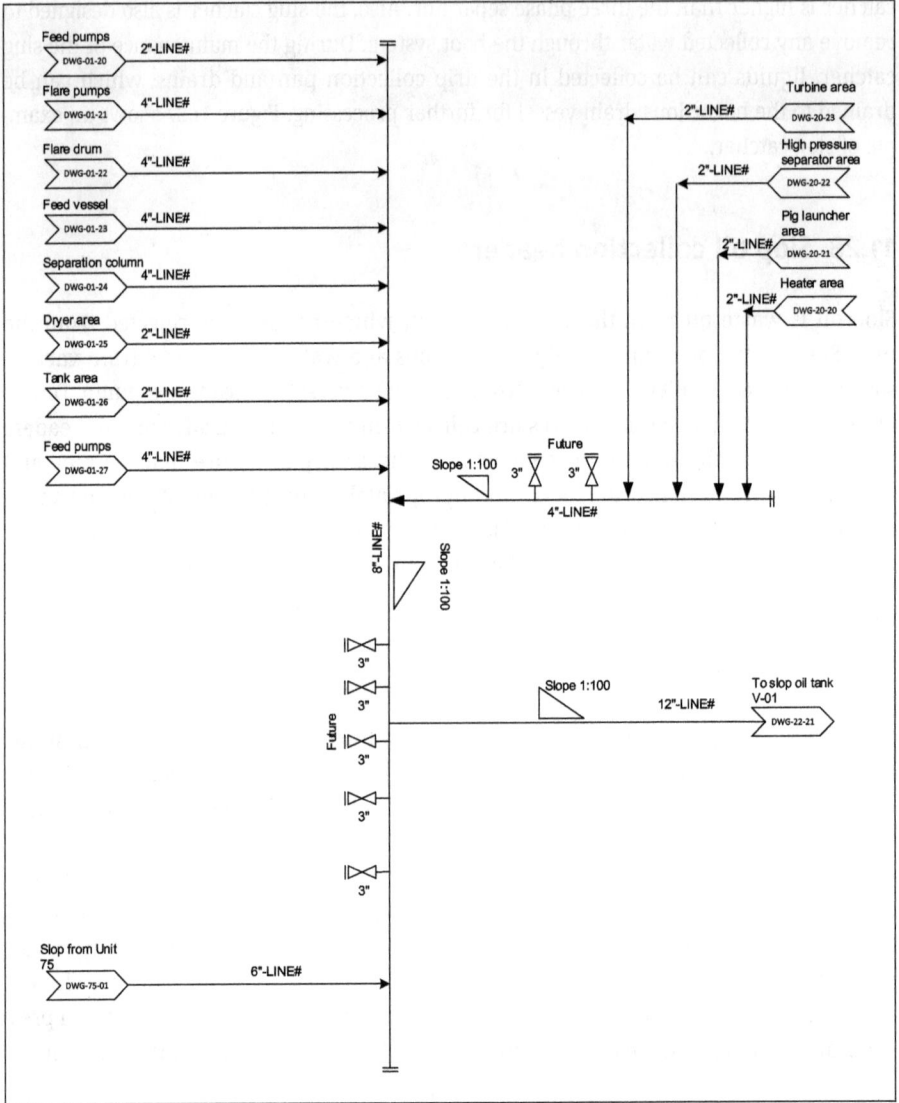

Figure 11.28: Slop oil collection header.

health of the filters. The level indicator on the sump can activate or start the pumps when the level is high. Alternatively, the process operator can manually start the pump by looking at the level inside the control panel. Figure 11.30 shows an example of such a sump system.

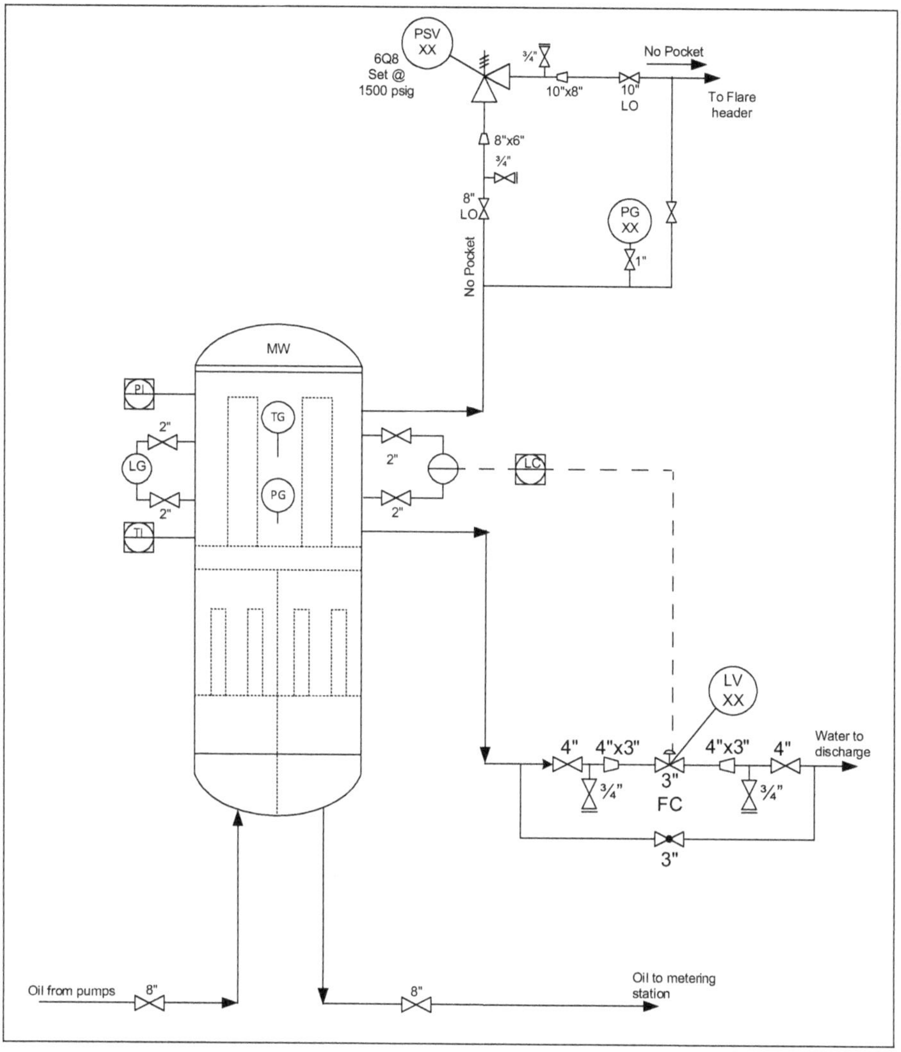

Figure 11.29: Liquid coalescer.

11.31 Fired heater

A fired heater is similar to an exchanger, but the heat transferred in the fired heater is based on convection and radiation mechanism. The cold fluid to be heated is passed through heater tubes in convection and radiation sections. The convection section is on the top of the heater, but the temperatures are relatively lower, compared to the radiation section. The radiation section is where most of the heat is transferred to the cold fluid in the firebox. The firing mechanism includes the burning of fuel gas or

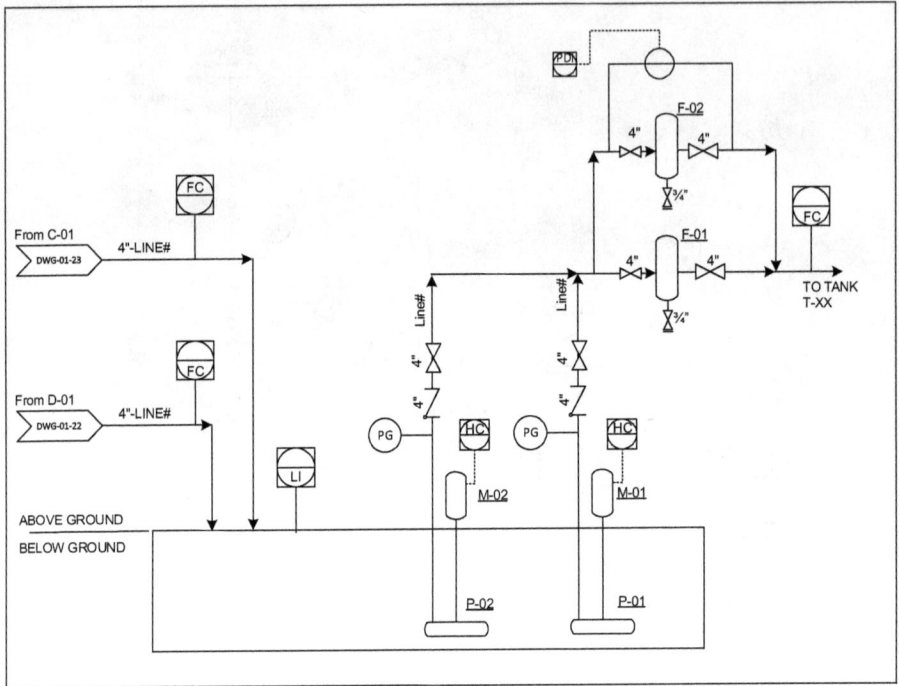

Figure 11.30: Sump system.

fuel oil. It is important to keep the minimum desired flow through the heater coil to avoid excessive heating of the tubes, which can lead to premature failure. Also, it is important to ensure that the flow through different passes is equal to avoid overheating one pass compared to another. In order to achieve that, typically, a flow control valve controlling the flow rate through that pass is often added. The temperature inside the convection and radiation section is monitored using multiple tube-skin temperature transmitters. The air required for burning can be received through a natural draft or a forced draft. If the heater duty is small, a natural draft air mechanism can be utilized. If the heater duty is large, a forced draft fan is needed to force the air inside the heater. It is important to keep an appropriate percentage of air inside the heater for adequate burning of hydrocarbons. In order to achieve appropriate burning of hydrocarbons, stack temperature and oxygen concentration are monitored. The draft on the heater is controlled using a draft mechanism installed on the heater stack. Figure 11.31 shows an example of such a fired heater system.

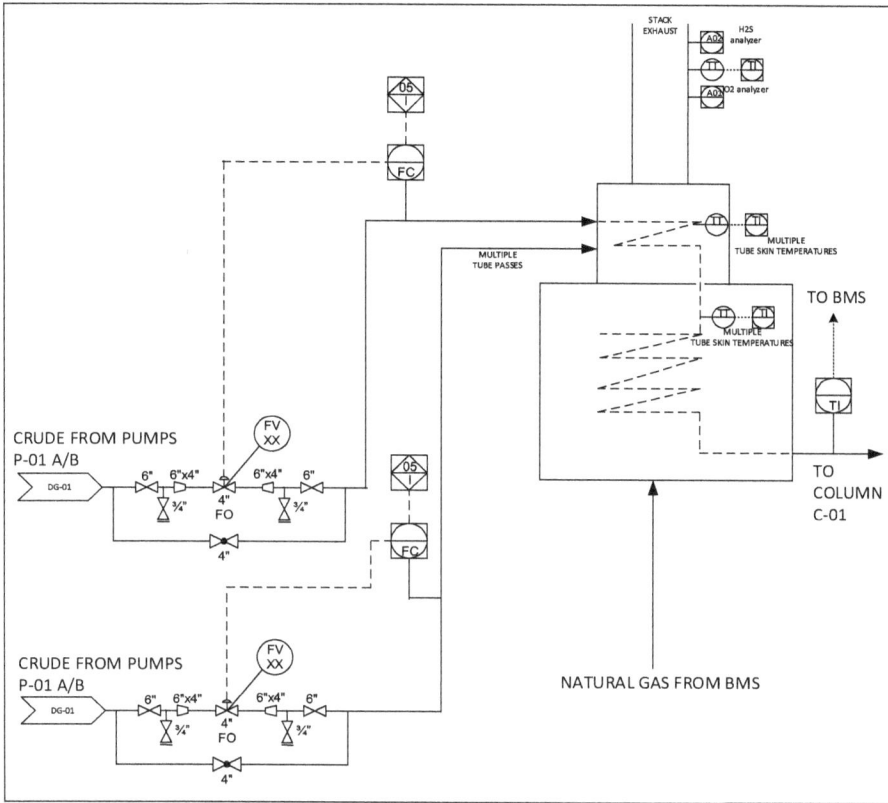

Figure 11.31: Fired heater arrangement.

11.32 Burner management system (BMS)

Natural gas is used as a heating medium for the fired heater in most of chemical and refineries plants. Natural gas comes in at high pressures of around 80 PSIG, and if the natural gas flow is not controlled, it can lead to a catastrophic failure of the fired heater. Natural gas from the supply is initially passed through a set of filter-coalescing elements, which removes any liquids. Liquid carried over to the fired heater is not desired as it can cause unsafe scenarios in the heater. Cleaned natural gas further splits into the main natural gas line, which is the main supply line to the heater, and a pilot gas line, which is used to keep the heater fire burning all the time at a very minimum firing rate. The continuous rate of pilot gas is important to make sure that the fired heater is not creating any unsafe scenarios by accumulating unburned hydrocarbons. The main natural gas line has flow control, which essentially regulates the flow of the natural gas, based on the heater outlet temperature. In addition, there are two shutdown valves on the main natural gas as well as the pilot gas line to isolate the natural gas lines from the heater, in case of an emergency. Any trapped natural gas between the two shutdown valves is vented to a safe location using a smaller shutdown valve. The shutdown of the heater is triggered by shutdown instruments such as high-high pressure and low-low pressure. High-high pressure indicates a malfunction in the natural gas controls, and high-pressure in the heater is not permitted as it can lead to catastrophic failure of the heater. Low-low pressure of the natural gas is also not permitted as it could turn off the firing in the heater, which can release unburned hydrocarbons into the atmosphere. On the pilot gas line, natural gas pressure is regulated by a mechanical pressure regulator valve to a fixed value, for example, 8 PSIG. Figure 11.32 shows an example of such a burner management system.

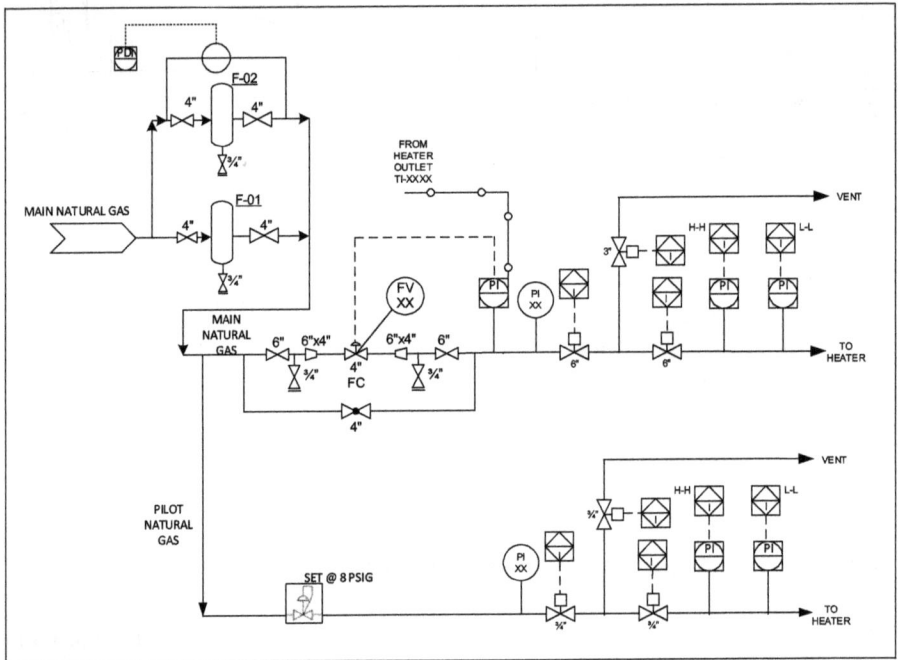

Figure 11.32: Burner management system.

11.33 Cooling water network

The cooling water network consists of a supply header, a return header, and several cooling water users. A set of cooling water pumps transfers cooling water to the supply header. Typically, in the chemical or refining plant, the cooling water supply and return headers are big in size, such as 30" or 42". Due to bigger pipe sizes, in addition to carrying liquid water, these supply and return headers are located either underground or just above the ground to avoid excessive load on the pipe rack and structure. The sub-header going to different users is taken from this supply header, and the return from the same user is sent back to the return header, as shown in Figure 11.33. If the cooling water system for a particular user develops a problem such as fouling or cleaning, the exchanger needs to be isolated from the other consumers. To achieve the isolation, isolation valves are added on the inlet and outlet of the exchangers. When designing the cooling water system, adding valves for future connections is always recommended.

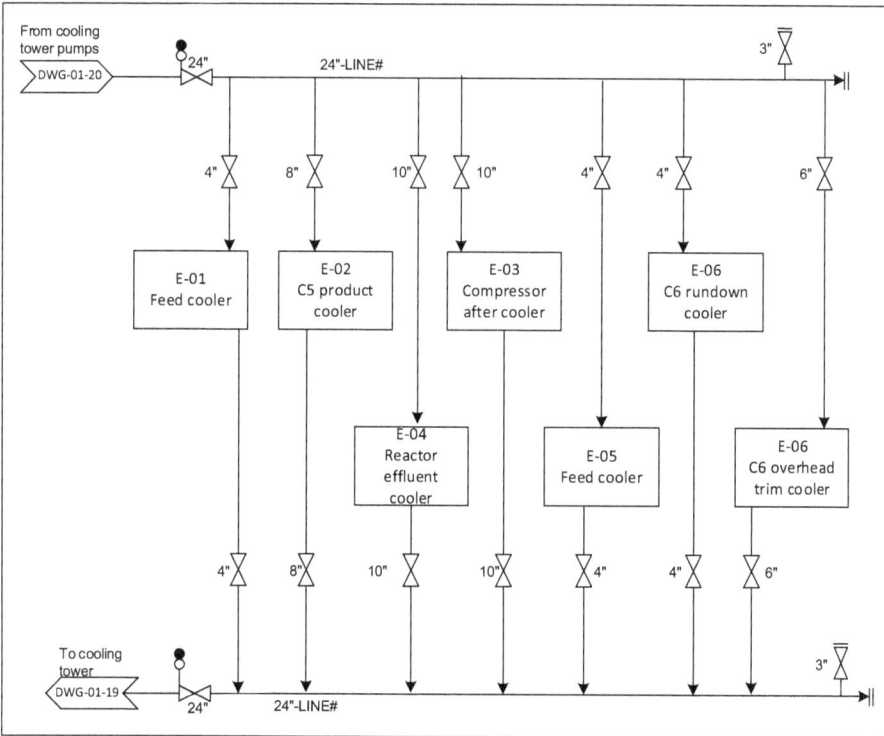

Figure 11.33: Cooling water network.

11.34 Vendor scope

On an engineering project, there could be more than one vendor package. During the early phases of the project, information on different vendor packages is unknown. An example of a vendor package is a compressor package, which involves the compressor itself, surge control, knockout suction drum, aftercooler, compressor seal system, control and logic piping, and instrumentation diagrams. The vendor scope is identified in a box with the anticipated number of vendor drawings mentioned inside. The necessary tie-ins are added from the vendor and to the vendor equipment. These tie-in numbers are used to communicate process development from the project engineering contractor to the vendor or vice versa. Typically, all vendors would like to know about the utilities available. In addition to the process streams, utilities can be shown with the line sizes. An example of the vendor scope drawing is shown in Figure 11.34.

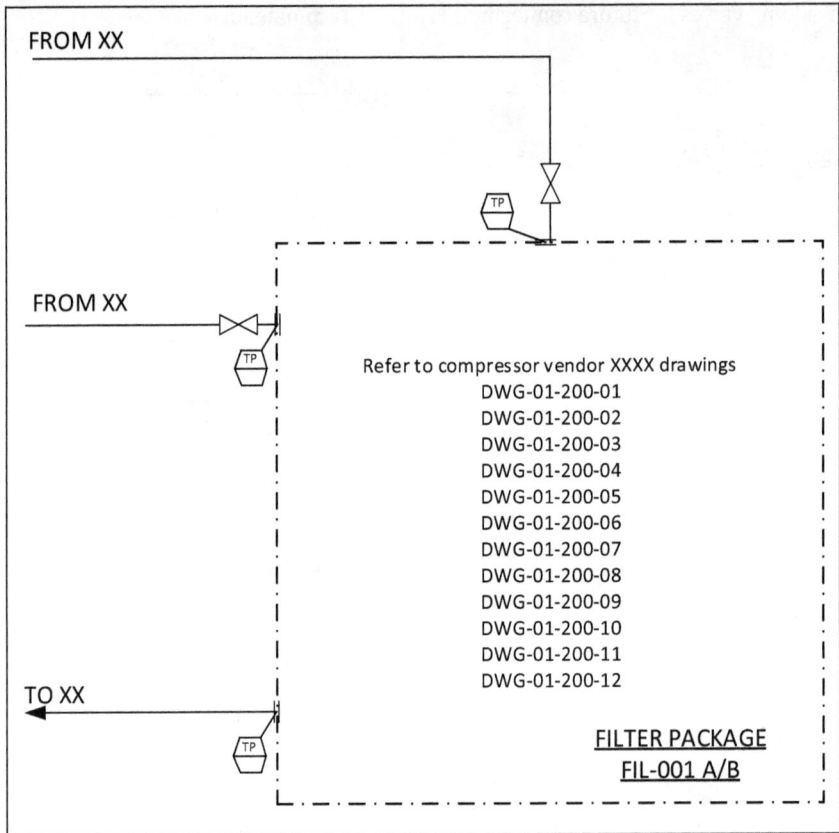

Figure 11.34: Vendor scope.

11.35 Battery limit drawing

A battery limit drawing is required for every unit in a processing plant. The battle limit drawing consists of a flow of a process and the utility piping to and from the unit. The drawing basically shows the connectivity of the unit with other units within the processing plant. The battery limit drawing shows the boundary between the unit and a tank farm area. A common instrument, such as flow, pressure, and temperatures, is installed on each pipe to quantify the flow and other process parameters. The battery limit lines also have block-and-drain valves with spectacle blinds. These help isolate the unit from the rest of the outside units during startup or shutdown events. Such a system is shown in Figure 11.35. Note that there could be multiple battery limit drawings, depending on how the pipelines enter and exit the unit at different locations within the unit.

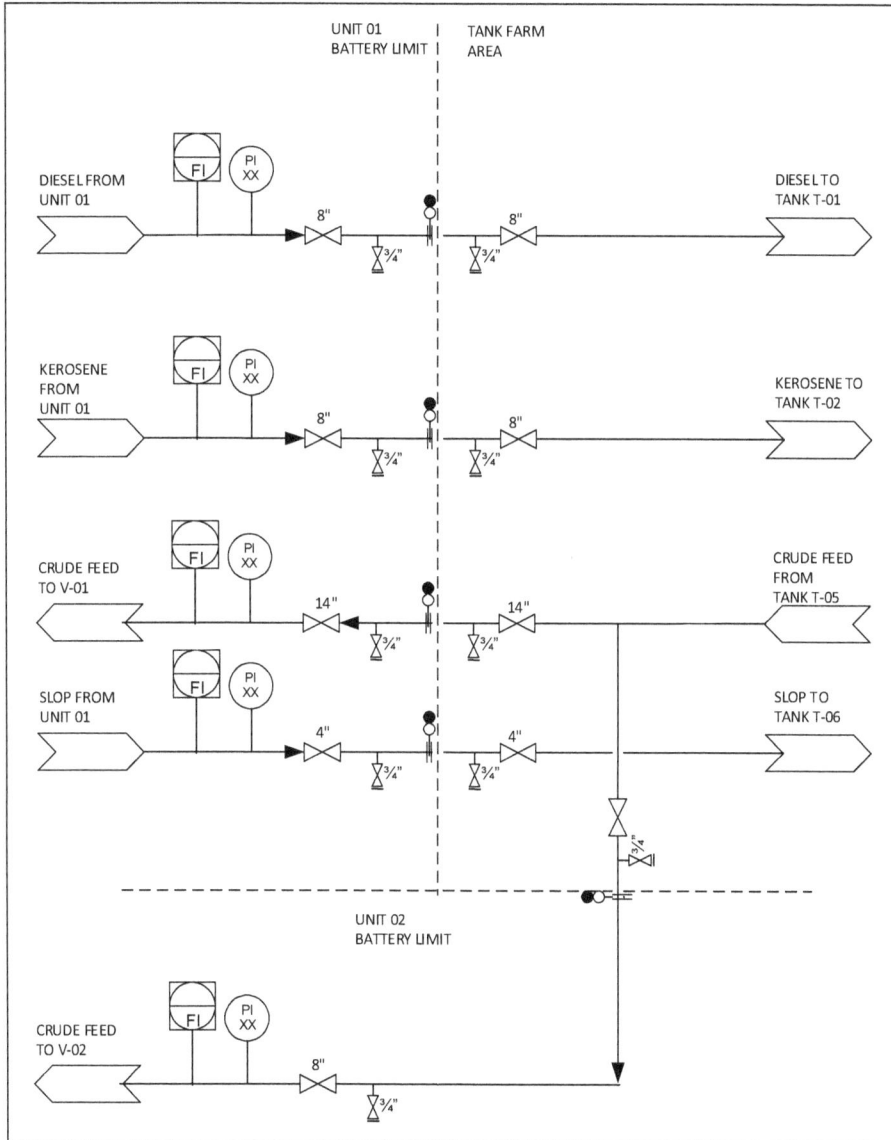

Figure 11.35: Battery limit drawing.

11.36 New scope vs. existing system

Most projects these days involve revamping existing units. In the revamp scope, a new piece of equipment or piping is installed, replacing the existing equipment or piping. The new scope of the project needs to be differentiated from the existing piece of piping and equipment so that all disciplines as also the customer can focus on the new scope. The new scope is usually denoted by a scope cloud, as shown in Figure 11.36. In Figure 11.36, the broken cloud before the 8" valve represents the as-built piping and equipment. A tie-in number and a solid cloud around the new piping show the changes done by the project.

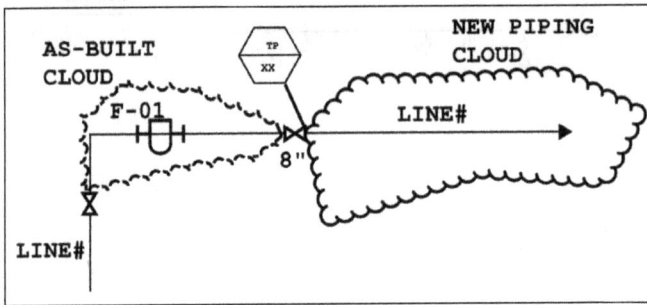

Figure 11.36: New scope versus as-built system.

Abbreviations

3D	Three-dimensional model
ACFM	Actual cubic feet pet minute
AEM	Type of shell and tube heat exchanger
AEL	Type of shell and tube heat exchanger
AEU	Type of shell and tube heat exchanger
ARC	Automatic recirculation valve
ARU	Amine recovery unit
AOP	Auxiliary oil pump
BFD	Block flow diagram
BFW	Boiler feed water
BHP	Brake horse power
BMS	Burner management system
C.S.	Carbon steel
C_{1-4}	Carbon chain of length 1 to 4
CSO	Car seal open
CSC	Car seal close
CO_2	Carbon dioxide gas
Cv	Characteristics of a control valve
CRU	Catalytic reforming unit
CDU	Crude distillation unit
DBB	Double block and bleed valve arrangement
DCS	Dynamic control system
DEMO	Demolition
DP	Differential pressure
DGA	Diglycol amine
EFDs	Engineering flow diagrams
ELDs	Engineering line diagrams
EPC	Engineering, procurement, and construction
ES	Emergency shutdown
ETR (ET)	Electrical tracing
FV	Full vacuum or flow valve
FI	Flow indicator
FC	Flow control or failed closed position
FCCU	Fluid catalytic cracking unit
FO	Failed open position
FAL	Low flow alarm
FALL	Low low flow alarm
FIP	Fire protection
GPM	Gallons per minute
HAZOP	Hazard and operability study
HP	Horse power
HC	Hydrocarbons
H_2S	Hydrogen sulfide gas
H_2SO_4	Sulfuric acid – chemical
HCl	Hydrochloric acid – chemical
HPV	High point vent
HEC	Heat conservation

https://doi.org/10.1515/9781501519864-012

ID	Inside diameter
IFA	Issued for approval
IFC	Issued for construction
IFD	Issued for design
IFE	Issued for an estimate
IFH	Issued for HAZOP
ITP	Current to pneumatic
K.O.	Knock out
LPG	Liquefied petroleum gas
LC	Level control
LT	Level transmitter
LO	Lock open
LC	Lock close
LP	Low pressure
LPD	Low point drain
MAG	Magnetic
MCC	Motor control center
MFD's	Mechanical flow diagrams
MMBtu/hr	Million metric British thermal unit per hour
MO	Motor
MOC	Material of construction
MDMT	Minimum design metal temperature
MOV	Motor operated valve
MTO	Material take-off
MDEA	Methyl diethyl amine
MMU	Master mark-up
NPSH	Net positive suction head
NOx	Nitrous oxide gas
NaOH	Sodium hydroxide – chemical
NFPA	National Fire Protection Association
NH_3	Ammonia gas
O.D.	Outside diameter
P&IDs	Piping and instrumentation diagrams
PFDs	Process flow diagrams
PSIG	Pounds square inch – gauge
PV	Pressure valve
PSV	Pressure safety valve
PC	Pressure control
PI	Pressure indicator
PDI	Pressure differential indicator
PG	Pressure gauge
PEP (PP)	Personal protection
PVRV	Pressure vacuum relief valve
PPM	Parts per millions
RAD	Radar
RTD	Resistance temperature detection
SP	Set point
SBA	Strong base anion
SRU	Sulfur recovery unit

STR (ST)	Steam tracing
SI	Specialty item
SWS	Sour water stripper
SGU	Saturated gas unit
TEMA	Tubular exchangers manufacturers association
TC	Temperature control
TG	Temperature gauge
TGTU	Tail gas treatment unit
TDH	Total developed head
TP	Tie-point
T/T	Tangent to tangent
VFD	Variable frequency drive
WBA	Weak base anion
UFDs	Utility flow diagrams
USG	US gallons
ULT	Ultrasonic
VDU	Vacuum distillation unit
XV/ SDV	Shutdown valve

Questions

Chapter 1

1. Who uses P&IDs and why?
2. How are P&IDs helpful in commissioning and hydrotesting activities?
3. Why are P&IDs so important for any process and plant operations?
4. When would an engineer start developing P&IDs?
5. What details would one expect from a P&ID?
6. Explain the differences between BFD, PFD, and P&ID?
7. What are the common tools used for the development of P&IDs?
8. What are some general thumb rules for developing P&IDs?
9. What are the types of P&IDs?
10. What are the general steps involved in developing P&IDs?

Chapter 2

1. Why should space requirements on the P&ID be understood and planned before marking any P&ID?
2. At which location would you show the design information of the storage tank and pump? Provide design details for a pressure vessel.
3. What is VFD control for a pump? How do you show the VFD control on a P&ID?
4. For liquid such as light naphtha, which drain system is recommended for a pump?
5. Why is a suction K.O. drum required for a compressor? Which parameters are important to monitor for the K.O. drum and why?
6. What are the typical internals of a three-phase pressure vessel, and what is the significance of each internal?
7. What are the typical instruments for a floating roof tank? Explain the purpose of each instrument?
8. Why is reactor temperature important, and what are the safeguards shown on the P&IDs?
9. What are common types of heat exchangers, and where are they used based on type?
10. Why is pressure control in a distillation tower important? What instruments are provided to make sure the pressure is controlled in the tower?
11. Why is superheated steam critical for the operation of the turbine?

Chapter 3

1. What is piping? Why does industry use piping?
2. What parameters are considered in pipeline numbering?
3. How do we determine line sizes?
4. How does one select piping specifications for a project?
5. What is the piping class? Why is it important?
6. Why are smart plant P&IDs effective for pipe design?
7. Why would you consider future expansion possibilities when designing piping?

https://doi.org/10.1515/9781501519864-013

Chapter 4

1. What are in-line and offline instruments?
2. What are the different types of in-line valve instruments, and how are they used?
3. What are different instruments available to measure a flow? Which instrument is most commonly used in the industry?
4. What are relief valves, and how are they classified?
5. Explain the function of pressure and temperature gauges.
6. How is the transmitter different than the gauge? What are the common types of transmitters?
7. What are common types of level transmitters, and where are they used?

Chapter 5

1. What are the types of controllers, and how do they differ in their functionality?
2. Provide an example of a cascade control loop.
3. Why are process alarms important, and how do they differ from safety valves?
4. What are cause and effect tables? Why are they used?
5. Provide an example of a safety interlock from any industry.

Chapter 6

1. What are utilities, and how are they different from process fluids?
2. Why do steam and condensate system have multiple levels, and how is the temperature controlled?
3. What are the most common applications of low-pressure nitrogen?
4. Why is maintaining 80–90 psig pressure for instrument air critical for plant operations?
5. Can you use plant air for the instrument air applications and vice versa? What are the implications?
6. How does the fire water tank maintain level all the time?
7. How is the BFW connected with the soft water? Can you produce BFW from clarified water instead of soft water?
8. Which ions are removed to produce demineralized quality water?
9. What is the concern with having H_2S in the fuel gas?
10. How is the correct hot oil fluid selection made?
11. What is common between tempered water and chilled water circuits?
12. Why is it critical to isolate utilities from process vessels or lines?
13. What are the different types of utility stations?
14. Why is the design of utilities so critical and an additional margin recommended?

Chapter 7

1. What are manual valves?
2. Which manual valves are to be used for controlling large and small amounts of flow?
3. Which valves have the lowest pressure drops?
4. What is DBB, and where does one use DBB arrangement?

5. Why are drain and vent valves necessary, and where are they provided?
6. What are the different types of blinds, and how are they used in the industry?

Chapter 8

1. What are P&ID legends? How are they helpful?
2. Why is personal protection insulation required?
3. What are some common utility services used in the industry?
4. Name five different equipment types and list their functions.
5. What types of strainers are used in the pump suction?
6. Why is it necessary to differentiate between different scope clouds?
7. What is the difference between a design note and a project note?
8. How are hold notes helpful for other disciplines?
9. Which location is preferred for a sample station?
10. Why are the lists of P&IDs, PSVs, control valves, etc. important?
11. Why is P&ID checklist needed? What is the outcome of the checklist?

Chapter 9

1. Why is it important to slope the flare lines?
2. Why is a flare seal system needed?
3. How is the deluge system different from the fire water system?
4. How do safety shower and eyewash stations help bring safety to the plant?

Chapter 10

1. What is the role of a P&ID drafting technician?
2. Who checks the drafted copies of P&IDs?
3. Why is MMU process needed, and who makes the MMU marks?
4. What are the different issues of P&IDs and what is their purpose?

Chapter 11

1. Why do control stations have drain valves, and why are they located at an accessible location?
2. How does a steam desuperheater works?
3. What type of steam trap is to be used for full recovery of the condensate?
4. What are the differences between liquid- and gas-type hazardous category relief valves?
5. Why are pockets not allowed in the relief outlet piping?
6. What are the criteria for using a flow restriction orifice for a pump design?
7. For a pump with a minimum circulation flow controlled using a control valve, what should be the fail-safe position of the control valve and why?

8. Suppose there is a trace amount of oil in the water, which drain or vent system does the water from a pump need to go to?
9. Why are pulsation dampeners required for a positive displacement pump?
10. How would you measure a pressure drop across a reactor system?
11. What are the different internals of a three-phase separator, and what is their significance?
12. Why is level control in the overhead column vessel important? How is the level controlled?
13. What is the impact of a very low level and fail opening of the liquid control valve for a compressor knockout pot?
14. Why is a hazardous open drain vessel vent system vented to the flare?
15. What is a tube rupture scenario for a heat exchanger? How does it affect the safety aspects of the exchanger?
16. What are the criteria for using air coolers? How does one control the duty of the air cooler, which has a fluctuating temperature in different seasons?
17. Why are double-pipe exchangers small?
18. How does a kettle-type exchanger differ from other types of exchangers?
19. What is the purpose of the ejector system?
20. How does one make hot water at a desired temperature using a utility station?
21. What is the function of tank mixers and heaters?
22. How do you prevent forming of a vortex in a distillation column?
23. What is the purpose of the pig launcher and receiver?
24. Why is a large volume provided for a slug catcher?
25. How to protect a liquid coalescer pressure vessel from an overpressure scenario?
26. How is the fired heater connected with the BMS?
27. What are some examples of piping network systems in a plant?
28. How does a vendor scope drawing help develop the project scope?
29. What information is provided for battery limit drawings?

Answers

Chapter 1

1. P&IDs are generally used by all disciplines in plant operations as well as by EPC contractors. P&IDs are used by field technicians, plant operators, contractors, and engineers to understand the process unit operation.
2. Hydro testing of a plant operation involves pressure testing of piping and several equipment. Commissioning of a plant involves starting a plant from complete shutdown. Both these events involve tremendous amount of planning and scheduling, and segregation of different areas in a plant system. The planning of commissioning and hydro testing can be achieved by using P&IDs.
3. P&IDs are critical in creating mechanical datasheets, process datasheets, control loop diagrams, instrumentation diagrams, construction planning, line list development, piping tie-in list development, piping isometric development, 3D model reviews, HAZOP meetings, and equipment design reviews.
4. P&IDs are only needed once someone has an idea to make a profit through manufacturing a product. Once the idea is fully wetted and approved by the management, the first step is to develop BFDs and PFDs. Once the BFDs and PFDs are developed and reviewed, P&ID work can be started.
5. P&ID can provide information regarding piping, equipment, the connectivity of different process equipment, notes related to design, revisions, drawing numbers, and service of equipment.
6. BFD is a basic block flow diagram showing the process's high-level scope. PFD shows more details regarding the connectivity of equipment and piping. P&ID shows sizing details of lines, equipment, and instruments.
7. Adobe is used during an initial scope planning phase. Bluebeam software is used to develop P&IDs in the define phase and detailed design phase. Autocad and Microstation software are used for drafting the P&IDs.
8. Correct spacing of equipment, not crossing lines, identification of hold items early, use of correct colors for marking, and correct use of pipe off-page connectors are some general thumb rules for developing P&IDs.
9. P&IDs are divided mainly by process, utility, auxiliary, tie-in, and demolition.
10. General steps include understanding P&IDs, locating equipment on P&IDs, connecting all equipment by pipelines, adding instrumentation details, adding necessary utility services, adding isolation, drain, and vent valves, adding notes, and developing safety system design details.

Chapter 2

1. Every equipment takes different space on a P&ID, depending on the complexity. If too many equipment are added to a single P&ID, it is hard to read and communicate the design information. As most projects are scheduled-based, any possibility of rework saves critical time for the project. By knowing the required space for equipment, P&ID rework can be avoided, which improves the efficiency of the P&ID development process.
2. Design information for a storage tank, since it is static equipment, should be shown on the top section of the P&ID. Design information for a pump, since it is a rotary equipment, should be shown on the bottom section of the P&ID. Design details of a pressure vessel is as shown below:

https://doi.org/10.1515/9781501519864-014

> Equipment tag: V–XX
> Name: Service description
>
> Size: XX ID x XX T/T
> Capacity: XX USG
> Design pressure: XX psig
> Design temperature: XX °F
> Material: material of construction

3. VFD is a variable frequency drive for a pump. It is used when a pump sees changes in flow and head routinely. VFD lowers the power consumption of a pump. VFD receives a pump discharge pressure as the target, and controls the motor speed.
4. Since light naphtha is a volatile and hazardous compound, the pump's drain should be routed to the flare.
5. Compressor K.O. drum is required to remove any carried-over liquid, and it essentially protects the compressor from a catastrophic failure. Liquid level and pressure drop across the demister pad are important parameters to look for. Flow to the compressor, which is connected to the surge control, is also a critical parameter observed.
6. There are several internals in a 3-phase pressure vessel. The demister pad removes all the liquid from the gas stream. The vortex breaker removes any possibility of forming vortexes. The weir plate helps in separating oil from water. Inlet-calming baffle reduces the speed of the incoming liquid and helps in the removal of liquids.
7. The typical instruments for a floating roof tank are temperature, level, pressure, and safety instruments. Temperature instrument measures the temperature of the liquid stored. Level instruments indicate the level in the tank. A pressure instrument helps maintain constant pressure inside the tank. Safety instruments, such as pressure vacuum and emergency hatch, are needed to protect the tank.
8. The reactor temperature is important because runaway reaction temperature can deactivate the catalyst or create an unsafe release of hazardous chemicals into the atmosphere. The safeguards are to trip the reactor system based on multiple temperature measurements.
9. Shell-and-tube, air-cooler, double-pipe, and kettle-type heat exchangers are commonly used in the industry. Shell-and-tube exchangers are mainly used for medium-duty applications. Air-coolers are used for higher duty and primarily vapor or gas services. -pipe exchangers are used for very small-duty applications. Kettle-type exchangers are used in applications where heat recovery is a possibility.
10. Pressure in the distillation tower helps in maintaining the vapor-liquid equilibrium and also in maintaining the product specification. The overhead column piping has a pressure controller, and the pressure in the overhead vessel is controlled using a control valve.
11. The turbine operates based on a high rotation of the rotor blades; any liquid carried over can damage the blades. Superheated steam ensures that no condensate enters the turbine, protecting the turbine from liquid hammering scenarios.

Chapter 3

1. Piping is a circular-shaped metallic pipe that carries process fluid or utilities. Industries use piping as they are proven efficient ways for fluid transportation.
2. Pipeline numbering includes the unit number, line size, pipe specification, thickness of the insulation, and the type of tracing.

3. Line size is based on fluid rate, properties of the fluid being handled, and design guidelines for a pipe. A higher fluid rate would need a larger pipe size. Higher velocity and pressure drop / 100 ft are not allowed per customer guidelines.
4. Piping specifications are developed for a project and may already be available for a customer. Pipe specification depends on the fluid type and pressure inside the pipe. Higher pressure inside the pipe needs a higher class pipe rating. Highly corrosive fluids need stainless steel or alloy material.
5. Pipe class is important as incorrect selection of a pipe class can cause catastrophic failure. Piping class refers to the pressure rating of a pipe. An increase in the temperature of the process fluid decreases the pressure rating of a pipe.
6. Equipment, piping, and instrumentation design information can be embedded in P&IDs using the smart plant. The piping isometric drawings can be smartly created using the smart plant P&IDs effectively.
7. Future expansion consideration during a project avoids any future tie-in or hot work. The installed valves can be used for future projects.

Chapter 4

1. In-line instruments are installed on the pipe and interact with the fluid inside the pipe. Offline instruments are installed external to the piping and do not interact with the flowing fluid.
2. Control valves are used to control flow inside a pipe. On-off valves are used to start or shut down flow through a pipe. Control valves with a hand-wheel design are used when manual operation and automated control action are desired. Motor-operated valves are used for operating larger-size valves remotely. Automatic recirculation valves are used as a substitute for a control valve in a minimum circulation of a pump. Backpressure and pressure-reducing manual valves are used to maintain a fixed pressure upstream or downstream of the manual valve.
3. Annubar, ultrasonic, vortex, flow orifice, Coriolis meter, magnetic flow meter, turbine flow meter, and venturi flow meter are different types of flow meters. Flow orifice-type meter is a commonly used meter in the industry.
4. Relief valve protects equipment, piping, or group of equipment and piping from an overpressure scenario. Spring-operated, pilot safety, rupture disc, rupture pin, and pressure and vacuum relief valves are common types of relief valves.
5. Pressure and temperature gauges are located on the field, measuring and indicating pressure and temperature locally.
6. Gauges indicate a process parameter on the field. Transmitters indicate a process parameter in the control room. Transmitters are pressure, differential pressure, flow, level, and temperature.
7. Common types of level transmitters are DP, magnetic, and radar. They are used to measure the level of the tank and vessel.

Chapter 5

1. A flow controller is used for controlling the flow rate. A temperature controller is used for controlling temperature. Pressure and level controllers are used for controlling pressure and level, respectively.
2. The tray temperature of a distillation column is controlled by manipulating the reflux flow rate.
3. Process alarms alert process operators before an unsafe process event were to happen. Process alarms do not protect the piping or process equipment from overpressure, but the safety valves do.

4. Cause-and-effect tables are used to achieve the maximum possible safety for a process plant. They are used to study a process plant's safety and interlock logic. The cause-and-effect table consists of different causes and their effects on the process plant.
5. A high liquid level in the compressor knockout pot or high vibrations in the compressor trips the compressor on an interlock logic table.

Chapter 6

1. Utilities are neither consumed nor used in the process operation, but they provide cooling, heating, fuel, cleaning, and maintaining some pressure. Process fluids interact with each other and always change in phase, pressure, and temperature.
2. Steam piping usually encounters heat loss, which generates condensate, when steam is distributed throughout the plant. For these reasons, steam is superheated and distributed. A lower level of steam is needed as different pressure levels of steam users are available in the plant. It is more economical to produce 600 PSIG of steam in the boiler than a lower-level steam. When the pressure is reduced through a steam letdown station, the lower steam level has a high degree of superheat, which is removed using desuperheaters. Condensate or boiler feed water is added.
3. Low-pressure nitrogen is most commonly used in the tank or pressure vessel blanketing and for utility station purposes.
4. The plant operation has several control valves and on-off valves, which are operated using an instrument air supply. An 80–90 PSIG instrument air supply is necessary to avoid an unplanned shutdown. If an instrument air supply is lost, these valves will not function, and the plant goes to a safe state.
5. Plant air has some level of moisture as it is not dried. Instrument air is completely dried using air driers. Plant air should not be used in the instrument air application as the water can create operational problems. Instrument air should not be used in plant air applications as the production of instrument air is slightly more expensive than plant air.
6. The supply of clarified water maintains the fire water tank level. The level control valve on the clarified water line controls the water rate to maintain the level in the fire water tank.
7. BFW is made using soft water. BFW cannot be produced using clarified water, as clarified water has solids and minerals, which can damage the boiler tubes.
8. Cations and anions are removed from water to produce demineralized water.
9. The H2S in fuel gas produces sulfur dioxide gas. This gas is not accepted as per environmental specifications.
10. Hot oil selection is based on fluid properties such as specific heat, heat capacity, thermal conductivity, and viscosity.
11. Tempered water and chilled water are closed systems.
12. Hazardous chemicals or molecules can backflow into the utility header, which can create a hazardous situation for other process systems.
13. Different types of utility stations are water-air-steam-nitrogen, steam-nitrogen, water-steam, and air-steam.
14. Utilities are required for all process plants and operations. A plant cannot run without them. An additional design margin on utilities is recommended due to future expansion considerations and changes in utility consumption.

Chapter 7

1. Manual valves are block valves that help in isolating piping or equipment.
2. Needle valves can be used for controlling small amounts of flow. Globe valves can be used for controlling large amounts of flow.
3. Butterfly and ball valves have the lowest pressure drops.
4. DBB is an arrangement of two block valves with drains. DBB is used in high-pressure and hazardous chemical service operations.
5. Drain and vent valves are necessary to drain and vent pipes or equipment. They are provided in all applications, such as pipelines and equipment.
6. Spacer, spectacle, and paddle blinds are the types of blinds. They are used to be positively isolative and create space to help with maintenance operations.

Chapter 8

1. P&ID legend sheets guide in the development of P&IDs. They are helpful in numbering units, numbering P&IDs and instruments, selecting process and utility services, valves and equipment, and selecting pipe fittings and clouds.
2. Pipe fluid with a temperature above 140 °F can cause harm to personnel. For such piping, it is necessary to add personnel protection.
3. Cooling water, utility air, instrument air, utility water, steam, fuel gas, and nitrogen are commonly used utilities in the industry.
4. Pump – used for transferring liquid, exchanger – used for heating or cooling, compressor – used for compressing gas, turbine – used for expansion of a gas, agitator – used for mixing fluids
5. A temporary (Y-type) strainer is typically used in the pump suction system to remove solids that could damage the pump internals.
6. Differentiation between two different scope clouds is necessary to help execute the project correctly.
7. A design note is added to convey an important design-related message to the reader. The project note is added to convey a temporary message to the other disciplines working on the project.
8. Hold notes represent insufficient information or incomplete design. These notes are helpful to alert other disciplines that the design is incomplete.
9. A sample station is preferred across a pump as there is a pressure differential across the pump.
10. List of P&IDs, control valve, and PSVs help to find equipment or piping during busy meetings.
11. The P&ID checklist is needed to check the quality of P&IDs. The auditor can provide improvement suggestions to the engineer who prepared the P&IDs.

Chapter 9

1. A slope of the flare line is needed to ensure the flare header does not accumulate any liquid. Any liquid accumulated can create a restriction to the relief path.
2. A flare seal is needed to avoid flame from the flare tip backflowing into the flare header network, which can create a potentially unsafe situation.
3. The fire water system is already pressured with the water. The deluge water header network system is not pressured, but it gets water on activation of fire or an unsafe event in the plant.

4. Safety showers and eye wash stations provide necessary clean water to wash off any chemicals from the personnel's body or eyes. This is critical as the affected person cannot run to a hospital immediately.

Chapter 10

1. The P&ID drafting technician drafts the P&ID marks, checks the drafted copy, and sends out the drafted copy to the P&ID drafting coordinator.
2. The drafted P&IDs are primarily checked by process chemical engineers and other discipline engineers, as needed.
3. MMU process is needed to avoid getting any marks made to the design in the detailed design engineering. The process chemical engineer and other discipline engineers mark the MMUs on the P&IDs.
4. IFE – Issued for an estimate is used for estimating the phase-2 cost. IFR – Issued for review is used for sending P&IDs for customer's review. IFA – Issued for approval is used to get approval from the customer. IFD – issued for design is used for completing the project's design phase. IFC – issued for construction is used to initiate the construction of a project.

Chapter 11

1. Control stations have drain valves so that the pressurized fluid can be drained for removing the control station for maintenance purposes. They are located at grade so that maintenance can be carried out easily.
2. Superheated steam is superheated using condensate or BFW. The water stream has a control valve that controls the water rate, based on the target temperature of the team.
3. Recovered condensate from this steam trap can be hard-piped and sent to the condensate recovery system.
4. There is no difference between the gas and liquid relief valve piping and valving setup as both relief valves go to the flare, and the flare line is sloped. Both relief valves are designed for different relief scenarios.
5. Pockets in the relief valve piping can create additional restrictions to the relief path.
6. A restriction orifice can be used for a pump minimum circulation line if the pump design falls under 50 gpm or 50 horsepower motor.
7. It should be a fail-open position as the control station protects the pump from minimum flow during an instrument air failure event.
8. The water with some oil drain from the pump should go to the open drain hazardous drain system.
9. They are required to stabilize the flow and ensure stable operation of the pump.
10. Pressure drop across a reactor is measured using a differential pressure transmitter.
11. The inlet-calming baffle reduces the speed of liquid and removes them. The separating baffle plate helps in separating oil and water. The demister pad helps in reducing any carryover of liquid droplets to the gas streams. The vortex breaker helps in avoiding the formation of vortexes.
12. Column overhead vessel provides necessary reflux for the tower and liquid suction for the product and reflux pumps. The level in the drum is controlled by manipulating the product storage flow rate.
13. A low level in the compressor knockout pot can allow high-pressure gas to escape into the low-pressure liquid system, potentially creating an overpressure situation.

14. The hazardous open drain system can have some dissolved hazardous gases, which cannot be vented to the atmosphere. These need to be burned in the flare.
15. Tube rupture scenario for an exchanger exists when an exchanger is not designed as per the 10/13th rule, i.e., the low-pressure-side design pressure needs to be at least 10/13th of the high-pressure-side. It affects the design pressure of the exchanger and may need a relief valve if the criteria are not met.
16. Air coolers are used for gas or vapor services that have higher duties. The temperature of the product stream can be maintained by controlling the speed of the fans.
17. Double-pipe exchangers are small because they are sized for small duty.
18. The kettle-type exchanger produces steam as a byproduct, which other exchangers do not.
19. The ejector system is installed to create vacuum for a distillation system.
20. Hot water can be made by mixing utility water and low-pressure steam.
21. Tank jet mixers help in mixing tank contents to keep the product at a uniform specification. A tank heater is used for heating tank contents to keep the liquid from freezing.
22. Add a vortex breaker.
23. A pig launcher and pig receiver are used for cleaning the pipe.
24. A large volume of slug catcher is provided to collect slugs of liquid from offshore wells during start-up events.
25. Add a safety relief valve.
26. The BMS controls the operation of the heater and shuts down the heater during any upset events.
27. Chilled water, hot oil, tempered, and cooling water systems.
28. Vendor scope is unknown for the length of time in a particular project phase. The vendor scope drawing has tie-ins identified, which help communicate with the vendor regarding possible developments from both sides, i.e., contractor and vendor.
29. The battery limit drawing has temperature, pressure, flow, valves, blinds, and flow directions.

Further reading

Developing P&IDs is a skill that can only be achieved by experience and knowledge of different process systems. Each system is different from another, but the chemical engineering fundamentals to develop the equipment design and P&IDs are more or less common. Also, the P&ID work differs significantly depending on the project phase. It is recommended to read the following references to enhance P&ID skills further so that the reader can contribute efficiently to their work assignments.

Cheremisinoff NP. Pressure Safety Design Practices for refinery and Chemical Operations [Internet]. Amazon. Noyes Publications Imprint; 1998 [cited 2023Mar28]. Available from: https://www.amazon.com/Pressure-Practices-Refinery-Chemical-Operations/dp/081551414X

Duncan TM, Reimer JA. Chemical Engineering Design and analysis An Introduction [Internet]. Google Books. Google; 1998 [cited 2023Mar28]. Available from: https://www.google.com/books/edition/Chemical_Engineering_Design_and_Analysis/8bqtAQAAQBAJ?hl=en

Gavin Towler, Sinnott R. Chemical Engineering Design Principles, Practice and Economics of Plant and Process Design [Internet]. ScienceDirect. 2012 [cited 2023Mar28]. Available from: https://www.sciencedirect.com/book/9780080966595/chemical-engineering-design

Karre AV. Managing engineering, procurement, construction, and commissioning projects: A Chemical Engineer's guide [Internet]. Wiley.com. 2022 [cited 2023Mar28]. Available from: https://www.wiley.com/en-ie/Managing+Engineering%2C+Procurement%2C+Construction%2C+and+Commissioning+Projects%3A+A+Chemical+Engineer%27s+Guide-p-9783527829736. https://doi.org/10.1002/9783527829729.ch8

Nnaji UP. Introduction to chemical engineering: For Chemical Engineers and students [Internet]. Amazon. John Wiley & Sons; 2019 [cited 2023Mar28]. Available from: https://www.amazon.com/Introduction-Chemical-Engineering-Uche-Nnaji/dp/1119592100

Person HS, Harry, Silla. Chemical Process Engineering: Design and Economics [Internet]. Taylor & Francis; 2003 [cited 2023Mar28]. Available from: https://www.taylorfrancis.com/books/mono/10.1201/9780203912454/chemical-process-engineering-harry-silla

Plant Design and Operations [Internet]. ScienceDirect. [cited 2023Mar28]. Available from: https://www.sciencedirect.com/book/9780128128831/plant-design-and-operations

https://doi.org/10.1515/9781501519864-015

Index

https://doi.org/10.1515/9781501519864-016